Make:
Props and Costume Armor
Create Realistic Science Fiction and Fantasy Weapons, Armor, and Accessories

Shawn Thorsson

MAKER MEDIA
SAN FRANCISCO, CA

Printed in Canada.

Published by Maker Media, Inc., 1160 Battery Street East, Suite 125, San Francisco, California 94111.

Maker Media books may be purchased for educational, business, or sales promotional use. Online editions are also available for most titles (safaribooksonline.com). For more information, contact our corporate/institutional sales department: 800-998-9938 or corporate@oreilly.com.

Publisher: Roger Stewart
Copy Editor: Elizabeth Campbell, Happenstance Type-O-Rama
Proofreader: Liz Welch, Happenstance Type-O-Rama
Interior Designer and Compositor: Maureen Forys, Happenstance Type-O-Rama
Cover Designer: Maureen Forys, Happenstance Type-O-Rama
Cover Images: Jason Babler
Cover Photographer: Michi Termo
Indexer: Valerie Perry, Happenstance Type-O-Rama

October 2016: First Edition
Revision History for the First Edition
2016-10-16 First Release

See oreilly.com/catalog/errata.csp?isbn=9781680450064 for release details.

Safari® Books Online

Safari Books Online is an on-demand digital library that delivers expert content in both book and video form from the world's leading authors in technology and business. Technology professionals, software developers, web designers, and business and creative professionals use Safari Books Online as their primary resource for research, problem solving, learning, and certification training. Safari Books Online offers a range of plans and pricing for enterprise, government, education, and individuals. Members have access to thousands of books, training videos, and prepublication manuscripts in one fully searchable database from publishers like O'Reilly Media, Prentice Hall Professional, Addison-Wesley Professional, Microsoft Press, Sams, Que, Peachpit Press, Focal Press, Cisco Press, John Wiley & Sons, Syngress, Morgan Kaufmann, IBM Redbooks, Packt, Adobe Press, FT Press, Apress, Manning, New Riders, McGraw-Hill, Jones & Bartlett, Course Technology, and hundreds more. For more information about Safari Books Online, please visit us online.

How to Contact Us

Please address comments and questions concerning this book to the publisher:

Maker Media, Inc.
1160 Battery Street East, Suite 125
San Francisco, CA 94111
877-306-6253 (in the United States or Canada)
707-639-1355 (international or local)

Maker Media unites, inspires, informs, and entertains a growing community of resourceful people who undertake amazing projects in their backyards, basements, and garages. Maker Media celebrates your right to tweak, hack, and bend any Technology to your will. The Maker Media audience continues to be a growing culture and community that believes in bettering ourselves, our environment, our educational system—our entire world. This is much more than an audience, it's a worldwide movement that Maker Media is leading. We call it the Maker Movement.

For more information about Maker Media, visit us online:

- Make: and Makezine.com: makezine.com

- Maker Faire: makerfaire.com

- Maker Shed: makershed.com

To comment or ask technical questions about this book, send email to bookquestions@oreilly.com.

CONTENTS

PREFACE

WHY MAKE PROPS?

The answer to that question is simple: Sometimes you want to possess something that doesn't even exist.

For most people, there have been plenty of times in their lives when they've been watching a movie or TV show, or playing a video game, and found themselves wishing they could be more like the characters on the screen. People who make up words for things call this form of dreaming *escapism*, and it's one of the main reasons these kinds of entertainment exist. No matter what shape daily life may take, people can take a break and be spirited away to whole different worlds that are more to their liking.

For most people, that's as far as it goes. They witness the end of the evildoers, the adventurer's spaceship returns home, and then they step back through the looking glass—or crawl back through the wardrobe—and leave every part of that other world behind. For most people, that's the end of the experience.

But not if you're a Maker.

For a Maker, the escape doesn't have to end when the credits roll or when the campaign is won. Unlike the mere mortals of this planet, a Maker can have souvenirs from those fantastic worlds.

Do you have a yearning to dress like the armor-clad superhero that just saved the human race from certain extinction? You're about to learn how a humble floor mat can be turned into that suit of armor.

Do you want to hang that barbarian marauder's particular battle-ax over your mantle? You're just a couple sheets of plastic and a few bits of lumber away from having it.

Want that galactic warrior's helmet for your bookshelf? A quick trip to the hardware store and you're on your way.

And what's more, if you're a Maker, when someone points to that just-right thing on your desk and asks where you got it, you can beam with pride and answer, "I made it."

You're about to embark on a journey that will give you a whole new way of looking at the world around you. By the end of this book you'll start seeing new uses for everything you can get your hands on, and you'll be able to have the kinds of things non-Makers only get to dream of.

Before Getting Started

Before you get to work, though, we need to take a moment to review safety procedures, the tools you'll need, and a few points about responsible behavior.

When learning new skills, and working with new tools and materials, it's always a good idea to be educated about the safety concerns involved. Many of the projects in this book will require the use of potentially dangerous tools and chemicals. Read the warning labels on everything. *Heed* the warning labels on everything.

Personal Protective Equipment (Be Safe)

There are some times in life when circumstances conspire against us to create dangerous, unhealthy situations. When you're engaged in a hobby, this should never be the case. There's absolutely no excuse for not taking adequate safety precautions while building your own prop and armor replicas. The staff at the local hospital's emergency room won't be impressed when you tell them you're there because you were in a rush to build your spaceman suit and didn't have time to avoid mangling your body in the process.

With that in mind . . .

Don't Destroy Your Lungs

One of the biggest concerns when it comes to most prop-making processes is respiratory protection. The paints and resins being used usually contain volatile organic chemicals and emit noxious fumes as they dry or cure. While it might be tempting to save a little money by skimping on respiratory safety equipment, keep this in mind: at the time of this writing, a decent organic vapor respirator will cost somewhere in the neighborhood of $40. That's 40 US dollars. A lung transplant can easily run somewhere in the neighborhood of half a million US dollars, if they're able to find a compatible donor, and you'll still have to take immunosuppressant drugs for the rest of your life in order to keep your own immune system from destroying the implanted organ and causing . . . Look, just buy the respirator.

In addition to making sure that the respirator is properly sized to form a good seal to your face, you'll also want to read the specifications of the filter cartridges that go with it. It turns out that after prolonged exposure to open air they will only be effective for a certain amount of time (typically something like 40 hours) and will need to be replaced with new cartridges. The cartridge life can be prolonged by placing it in an airtight container when not in use, but once it's removed from the package, keep track of how many hours it's been in use. It's your health. Don't damage it by trying to save a few dollars.

Even if there aren't any chemical fumes to worry about, sanding, cutting, and grinding will often generate fine dust that can cause a lot of heartache if it's inhaled. This is when a cheap dust mask is a good idea. This is the *only* time it's a good idea.

Don't Destroy Your Eyes

Eyes are delicate things. When you're mixing chemical goo it's really easy to splatter some tiny drip of caustic ick into them and ruin your entire day. The same goes for having the tip snap off of a hobby knife and pop itself into your cornea or having some cloud of dust fill your eyelids with abrasive garbage. It's just something that's going to happen. Since you know it's going to happen, pick up a pair of safety glasses and wear them. You'll work a lot faster if you're not stopping to flush gunk out of your eyes with running water.

The pirate look may be in vogue lately and some people might consider an eyepatch to be a tasteful fashion accessory. But do you know what's even cooler than wearing an eye patch? Not answering the, "What happened to your eye?" question with some version of "I'm an idiot."

Don't Destroy Your Ears

At some point, power tools will be involved. If you're using a machine that requires you to raise your voice above a normal conversational volume *at all* in order to be heard, you should be wearing some form of hearing protection. There are countless options available. Pick something comfortable.

Don't Destroy Your Hands

Rubber exam gloves are almost always a good idea. Many of the materials you'll be using are harmful to your skin, but even if they're not, it's often nice to be able to simply peel off a layer of rubber and have relatively clean hands. That way you can answer your phone without gluing your fingertips to it.

Some people are allergic to latex. If you're one of those people, use nitrile gloves instead. You shouldn't have to be told.

If you're wearing work gloves to protect your dainty fingers and supple palms from the abrasive effects of sanding and polishing, make sure you wear gloves that fit properly and allow your hands to move freely. Better yet, grow some calluses and limit the glove-wearing to those occasions where things are hot enough to potentially cause burns.

FIGURE P-1: Fashionable protective equipment is an absolute must.

🔫 Maker Note

When selecting your safety equipment, you can get all kinds of advice about the best brands, the best types, and the best prices. Listen to all of it. Make sure your personal protective gear is adequate for the job it needs to do. What most people won't mention is that, all other things being equal, you should pick safety equipment that looks cool (Figure P-1). Face it, if you look silly in your respirator and safety glasses, you'll be less willing to wear them. However, when you find a setup that looks great on you, remember to take it back off when you're done working. Nobody at the nearby restaurant, supermarket, or airport will understand why you're wearing your safety gear and you'll be hard-pressed to explain it to the police or mental health professionals you'll eventually meet along the way.

Tools (Be Smart)

Having a state-of-the-art workshop filled with high-end gear and specialized equipment can be nice, but great-looking props can be made with just a handful of simple tools. If you're someone who took all of the shop classes in high school or you have a solid grounding in do-it-yourself projects, you may want to skip this section. But since we all didn't grow up tinkering in woodshops or building our own cars, there are a few basic things worth going over.

Knives

These are simultaneously the most indispensable and most dangerous tools that every workshop must have. It's ideal to have a heavy-duty utility knife (often called a *sheetrock knife*) with disposable blades, as well as a small hobby knife (such as an X-ACTO) for finer work (Figure P-2). When you're using a knife, remember to cut away from yourself (and those around you) and, when cutting through something, make sure you're cutting on top of a cutting board or some other surface that won't mind a little extra love from a sharp blade.

FIGURE P-2: A heavy-duty utility knife and a small hobby knife. Looking sharp!

Maker Note

It may seem counterintuitive, but the sharper a knife is, the safer it is. A dull blade will cause you to push harder to cut through things. This usually means there's less control involved and you're more likely to slip and cut something you don't want to cut—something like a major artery.

Sandpaper

Sometime shortly after the ancient Chinese first invented paper, some enterprising genius found a way to glue sand to it and start smoothing things out. It took centuries after that before someone else figured out a way to quantify the precise roughness of the sand glued to the paper so we could all benefit from a standardized system of numeric abrasive values known as grit. When purchasing sandpaper, the higher the grit, the smoother it is. Why is that? When sorting out the abrasive particles, they're poured through a mesh. The grit value is the number of holes along each edge of a one-inch square of that mesh. So 80-grit sandpaper is covered in particles that are no larger than 1/80th of an inch in size while 400-grit sandpaper is covered in particles that are no larger than 1/400th of an inch in size. Now you know.

There are a few key things to know about using sandpaper. If you're sanding something that's supposed to be flat, it's a good idea to use a sanding block (several are shown in Figure P-3). This can either be a purpose-made tool with an ergonomic shape and a clamping system for holding a piece of sandpaper onto it, or it can simply be a scrap of wood with sandpaper wrapped around it. The key feature is that it has a smooth, flat surface on one side to prevent you from applying too much pressure in one spot and digging out a groove or divot.

There are also plenty of times when the surface to be sanded isn't flat at all. For curved areas, it's better to sand with the paper folded and held in the hand. It turns out there are a lot of ways to do this wrong. Some folks will simply fold the paper in half and then fold it in half again. This leaves two of the four faces rubbing against each other, grinding away due to their mutual abrasiveness and eventually turning even the finest sandpaper into some really poor-quality notepaper.

The better option is to actually tear the paper in half short-wise, then fold the half sheet into thirds (Figure P-4). This way, the front and back of the piece of paper can be used for sanding, while simultaneously protecting the third segment as a reserve element to be called into action when the going is too rough.

FIGURE P-3: Various sanding blocks

Heat Guns

These are typically used to cook paint off of the side of a building in preparation for refinishing. They look and act a lot like an industrial-strength hair dryer. DO NOT USE ONE TO DRY YOUR HAIR! The thing gets hot enough to peel paint off of the walls for goodness' sake! Keep that in mind before grabbing some piece of plastic that's just been heated up to near-melting with a heat gun. It'll be hot. Wear gloves and be careful. Most

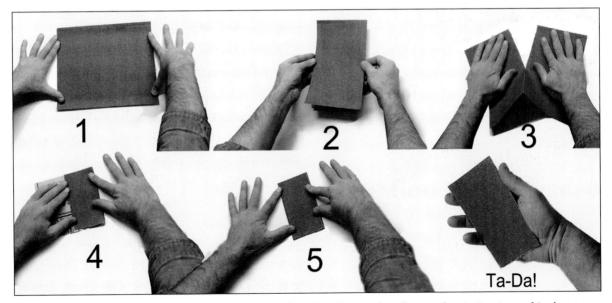

FIGURE P-4: This is the proper way to use sandpaper when hand sanding. Anyone who tells you otherwise is not your friend.

importantly, when the heat gun is not in use, make sure it is turned off and unplugged. That way it won't burn the whole house down.

Drills, Saws, Sanders, and All Manner of Scary Power Tools

If you see some sharp, spinning, motorized thing, and you don't know how it works, *ask* someone who does! Get him or her to show you how to use it safely before you lose a limb.

Don't know anyone who is familiar with power tools? Look for resources in your local area. Some of the big-box hardware and home improvement stores offer workshops and classes. Other hardware stores may offer consultation services. These days, in many urban areas, you can find *makerspaces* (also known as *fablabs*, *hackerspaces*, and *DIY spaces*) where people get together to share their tools and expertise. There are also companies such as TechShop that provide these services for a fee. If you can't find in-person help, you can always turn to books and online videos for instruction.

The point is, don't start using power tools without getting some kind of training. If you still think you'd rather just pick up a potentially dangerous piece of equipment and just start pushing buttons until you've figured out how it works, please put down this book and go back to eating paste before you hurt yourself.

Behavior (Be Cool)

Prop- and costume-making is a great way to pay homage to a character or story that you love. Once you've finished your creation, you'll want to show it off. In a way, it gives you an opportunity to be a sort of ambassador, sharing your knowledge and enthusiasm with those around you. Here are a few dos and don'ts to keep in mind when you're out in public:

DO RESPECT THOSE AROUND YOU. When trooping around or posing for photos in a crowded or public place, don't block or obstruct traffic. Be courteous. Respect people's personal space.

DO BE ESPECIALLY NICE TO CHILDREN. To some of the kids you encounter, you're not just some person in a homemade costume. To them, you *are* the character you're dressed as. Remember that when they see you and start getting excited. It's one thing if they kick you or abuse you, but otherwise, be the hero they think you are.

DO BE PATIENT WITH PEOPLE. At first, it'll be a lot of fun answering all of the questions and getting people's attention. They'll ask, "Where did you get that?" or "What's it made of?" or make some sort of silly joke about the character you're portraying (or something they've mistaken you for). Inevitably, it'll get old. For the thousandth time someone will ask the same, tired question. But no matter how many times you hear a particular question or joke, to the person saying it, it's their first time. So unless they start by being rude or unpleasant to you, be nice to people who are curious about your work.

You'll also have your painstakingly crafted costume mistaken for all manner of things it's not supposed to be. Remember, not everyone shares your enthusiasm for obscure science fiction films from Turkmenistan (or whatever it is you're into). When some uninitiated passer-by calls you "something from Star Trek," just smile and nod. Getting indignant won't do you any good.

DO MAINTAIN SITUATIONAL AWARENESS. A costume that has a full face-covering mask or helmet isn't something you should wear to the bank. If you forget this, the security guards or local police will likely remind you. It won't be fun. Also, if your costume has controversial or potentially offensive themes, it may not be the best choice to wear it out in public.

DO OBEY THE POLICE. When you're walking along the sidewalk in your badass science fiction warfighter outfit, you'll look pretty cool. But it's also quite possible that someone who sees the wooden plasma rifle slung across your back or the rubber blaster pistol strapped to your thigh will think it's all real-looking enough to call the police. If you do end up being stopped by law enforcement officers, don't give them a reason to fear for their safety. Remember that they have been called to the scene expecting to find a strange, armed person and they'll be jumpy. Follow their instructions with slow, deliberate movement and explain to them that you are not carrying any real weapons, only theatrical props that are not dangerous.

DON'T FRIGHTEN THE NORMAL. Simply wearing a costume in public shouldn't be a problem, but it may make a lot of people wary of you. Acting outrageously and being obnoxious is a bad idea. Don't do this.

DON'T THREATEN PEOPLE, EVEN AS A JOKE. You can never know how someone's going to react. It's not nice to frighten people who have no idea what's going on. Don't expect total strangers to happily engage in role-play they don't understand or don't want to play along with. Put yourself in their shoes. Imagine if you were out in public, minding your own business, and an armed, masked lunatic started waving some sort of weapon at you. Your first instinct probably wouldn't be to assume the weapon is a fake. Keep those sorts of shenanigans for a controlled environment.

DON'T HIDE BEHIND THE COSTUME. Dressing up as a villain doesn't make it okay to mistreat people. Playing with your friends is fine. Harassing random strangers is not!

DON'T BEHAVE BADLY IN GENERAL Going to a venue or event in costume and acting like a fool is NOT cool! Bad behavior on the part of costumed participants may well result in the venue or event in question banning costumes altogether. Don't be that jerk.

What to Expect from This Book

The following chapters will provide a solid jumping off point on the journey toward creating all manner of phenomenal props and costume pieces. That said, it will not be the end-all, be-all tome that will tell you everything there is to know about this amazing hobby.

Every experienced prop maker will have his or her own approach to solving every problem they encounter. If you ask a dozen Hollywood special effects artists how to create a particular effect, you'll probably get a dozen different answers, and they'll probably all work. The real trick is finding a way to make things that works with your abilities and the tools you have available.

Most importantly, this is not a cookbook. You'll learn quite a few processes as you read through it, but there's no step-by-step guide to help you just replicate a particular costume. As a matter of fact, each of the processes demonstrated in these chapters can be used to make any of the different items shown. Instead, think of it as an introduction; an invitation to take your first steps into an amazing new world.

What to Expect from Yourself

Effort.

The only difference between a dream and a goal is a plan. A plan plus effort becomes an achievement. You will need to put in some work. Whether you're assembling a Pepakura model by hand or carving shapes out of foam, there will be labor involved. Labor takes time, so you must make some time.

Your first prop replica may look like a complete abomination. Be ready for it. You'll get to the end of a weeks-long build, step back, and realize that your first attempt at replicating a galactic stormtrooper looks more like something that's been through the trash compactor. Everybody begins somewhere. Just know that your next project will be always be better than the last one, and keep building.

Finally, be aware that while you may be the most perfectly talented, capable Maker in the world, you will still make mistakes. Sometimes everything works out just the way you envision it. Other times, you get to learn something new. Those are the only two options at the end of any project.

Okay, we've gotten all the necessary lessons and lectures out of the way. So, what are you waiting for? Let's get started . . .

ACKNOWLEDGEMENTS

WRITING BOOKS IS HARD. Writing this book was long and hard. So long and so hard, I couldn't handle it on my own. With that in mind, I'd like to thank some folks in particular (in no meaningful order).

My editor/publisher Roger Stewart for his seemingly endless patience despite all of my delays.

Elizabeth Campbell, my other editor, who deleted all of my F-bombs and made it look like I know how to write good.

Jason Babler, one of my greatest patrons, for pointing Roger my way when he was looking for someone to write "a book on creating fantasy and science fiction props . . . a how-to book for people who are interested in trying to create cool stuff like you make." An amazingly talented sculptor and graphic designer, Jason also created the front cover image for this book.

My friend Matthew Herman for countless hours of helping me tinker in the workshop, as well as planting the seed of madness that has grown into years of making props and costumes and all manner of awesomeness.

My girlfriend Shawnon Kaiser for all of her encouragement, and her willingness to wear the skimpy wolf girl costume out into the snow in Upstate New York so we could get the one just-right photo.

Mandy Valin, Renaissance woman, for braving the same snowy woods to get location photos.

My new favorite photographer, Michi Termo, for setting aside a Saturday morning in her studio so we could shoot the cover.

Peter Rubin, my good friend and Hollywood concept artist, for coming up with the design for the Wolf Girl's armor despite his busy schedule.

Sarah Madill and Michael Bettencourt for giving up that same Saturday morning and strapping on the Wolf Girl and Hunter costumes for the cover shoot.

Anna Van Zuuk for getting dolled up and showing us how to vacform in the kitchen, and Lisa Bacon for letting us melt plastic in her oven.

Cary Gunnar Lee for his excellent rendering of Hunter as DaVinci's Vitruvian Man.

Alexander Belfor, comic creator, for letting me bring his character Hunter to life to illustrate many of my methods.

Binky Thorsson, my mother, who has supported all of my seemingly idiotic ambitions despite the fact that there's clearly no future in it.

Barry Thorsson, my father, who always manages to stop by the workshop and tell me the easier, simpler way to do something I've just wasted hours getting wrong.

My whole crew at Thorsson & Associates Workshop for listening to me work out sentences and try out diction to see what passes for funny outside the confines of my own head.

And finally, I'd like to thank all of the spineless little weaklings I've had to crush to get where I am today.

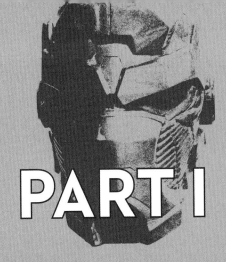

PART I

Prototypes and One-Offs

BUILDING WITH SHEETS AND TUBES
Starting Off Easy

A LOT OF THE things we need to make are just plain flat. When early cavemen were making theatrical props, this would cause all sorts of problems because building a perfectly flat surface from scratch can be difficult. Fortunately for modern-day prop builders, it's just a matter of buying flat stock and cutting out the necessary pieces. By stacking multiple layers, often in different thicknesses, you can make all manner of things. It's only a matter of identifying the different layers needed and getting started.

But that's not all. Somewhere along the way, the ancient Romans were struggling to build their own replica prop rifles and thought, "It's awfully tough to carve marble into the shape of a gun barrel." Then some enterprising genius invented lead pipes.* Then, tired of having everything taste like lead, some other genius decided that their massive stockpile of polyvinyl chloride blocks could be put to better use, and started melting them down and extruding the world's first PVC pipes.**

Once someone figured out how to cultivate the first orchards of dowel trees,*** with their perfectly cylindrical fruits, life became a dream for beginners in the realm of prop making.

---

*No, the ancient Romans weren't *really* trying to make rifle props. They were strictly pistol users.

**This isn't *actually* how PVC plumbing was invented.

***Dowels don't really grow on trees. Um . . . You know what we mean.

Starting Off Simple

To show just how easy it is to turn flat, sheet materials into all manner of awesomeness, let's take a look at a simple piece belonging to the character shown in Figure 1-1.

This guy has all sorts of gear that will be great to construct from flat materials, but to keep things as simple as possible in the beginning, let's start with Hunter's belt buckle (Figure 1-2).

Before jumping into a project like this, it's important to gather as much reference material as possible. Depending on the project, your reference material may come from graphic novels, they may be printed screen captures taken from movies or video games, or they may even be images saved from web searches. Get as much relevant reference material as possible and study it at length before beginning.

It helps to set up the workspace so that the reference images are easily visible without having to dig them out of a drawer and leaf through them with sticky, gloved fingers. Figure 1-3 shows a decent setup that keeps everything visible and off of the workbench.

The belt buckle prototype will be made out of stacked layers of medium-density fiberboard and various pieces of sheet plastic.

FIGURE 1-1: Hunter, a comic book character

FIGURE 1-2: Everybody's got to start somewhere. For this character, we will start right in the middle.

FIGURE 1-3: Taping the reference artwork to a piece of poster board can be really helpful.

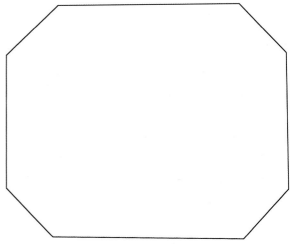

FIGURE 1-4: The outline of the belt buckle

Medium-density fiberboard, commonly referred to as *MDF*, is an inexpensive building material available in various sizes at most home improvement stores. It's essentially the sawdust swept up off the floor at the lumber mill, then pressed into a board shape with an adhesive to bind it all together.

⚡ Warning

The adhesive used to bind the sawdust together contains formaldehyde and all sorts of other carcinogenic compounds. This will be the first of many times you'll read the words *wear a respirator*, as well as *work in a well-ventilated area*.

The first step in making a piece like this is to determine the overall outline of the part. It helps to draw this out with the help of some CAD software (if you're that kind of person) or, if you're normal people, a straight edge (Figure 1-4).

Once the outside shape is determined, identify the different levels of the part (Figure 1-5).

Now, cut the outline shape out of a piece of MDF, as shown in Figure 1-6.

Next, grab some thinner material to make the other layers. In this case, the top and bottom sections of the belt buckle call for 1/8"- and 1/16"-thick plastic sheet, respectively.

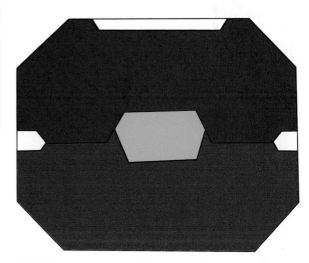

FIGURE 1-5: Each color is a different height.

FIGURE 1-6: The foundation layer

FIGURE 1-8: The belt buckle beginning to take shape

After drawing the outlines for the next two layers, transfer the shapes to the sheet plastic and cut them out. While it's simple enough to cut through thinner sheets with a sharp utility knife or a pair of scissors, thicker material may call for a rotary tool, a band saw, or a scroll saw, as shown in Figure 1-7.

Once each layer is cut out, it's simply a matter of gluing each one to the base layer as shown in Figure 1-8.

Hobby stores that sell plastic model kits and supplies for model train layouts will usually carry a wide variety of sheet styrene that includes various surface textures (Figure 1-9).

For the button in the middle of this belt buckle, a small piece of textured styrene helps it stand out as being something functional (Figure 1-10).

FIGURE 1-7: A scroll saw is ideal for making detailed cuts and even tightly curved shapes out of thin sheet stock.

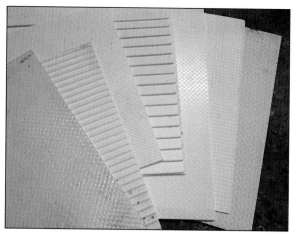

FIGURE 1-9: Textured styrene sheet intended to look like miniature siding, diamond plate, and roofing materials for tiny buildings.

FIGURE 1-10: The button

Now that the whole thing is assembled, it will need a coat of primer to blend the various bits of plastic into a single, uniform piece, as shown in Figure 1-11.

Once the primer is dry, it may be necessary to help blend the edges of the layers together using a bit of spot putty. This is an air-drying product that comes out of a tube like toothpaste. It can be found at most hardware or auto supply stores where it will be called something like *glazing and spot putty*, but hobby or craft stores will also carry a similar

product called *filler putty*, *green putty*, or *white putty*. In the end, they're all designed to fill up small holes or defects in a surface. Simply smear the putty into place with a fingertip, tongue depressor, or the tip of a hobby knife (Figure 1-12).

Once the putty has dried, it's simply a matter of sanding it down with progressively finer grits of sandpaper. Start with 150-grit sandpaper to do the initial smoothing, and then finish with 220-grit.

After sanding the buckle smooth, the next step is to give it a believable paint job, as shown in Figure 1-13.

How was it painted? That'll be covered in a later chapter.

FIGURE 1-12: The edges coated with a smidgeon of spot putty

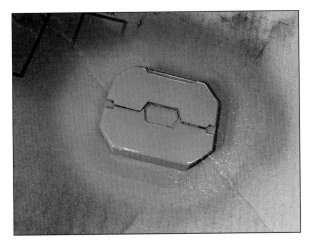

FIGURE 1-11: A coat of gray primer to show where all of the flaws are

FIGURE 1-13: Shiny!

Battle Axe for the Wolf Warrior

The belt buckle was a very simple project aimed at familiarizing you with the concepts. Now, for an example of how this same process can be applied to something a bit more impressive, take a look at the as-yet-unnamed warrior woman in Figure 1-14.

To give her weapon the right sense of thickness, the blade will be made out of two layers of ¾" (about 19 mm) MDF. The whole process begins with drawing out the profile of half of the double-headed blade onto a piece of MDF, as shown in Figure 1-15.

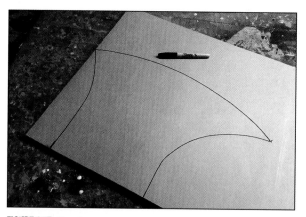

FIGURE 1-15: The shape of half the blade drawn out on the sheet

Then, using a jigsaw, cut out four copies of this profile (Figure 1-16).

Since the whole head of the axe will have to fit onto a shaft of some sort, the blade halves should be notched in order to fit cross pieces to tie them all together. The four matching cross pieces can be seen in Figure 1-17.

At this point, the two halves of each side of the blade need to be glued together. In order to keep everything from shifting around while the glue dries, clamp them together with spring clamps, or simply screw them together with some drywall screws.

FIGURE 1-14: We can call her *Ermahgerd, Warrior Maiden of the Wolf Clan* and her great-axe *Hedwak*.

FIGURE 1-16: Then shalt thou count to four. No more. No less.

FIGURE 1-17: These widgets will make a lot more sense in a moment.

Once the glue is completely dry, it's time to put an edge on the blade. This can be done by grinding it off with a coarse file, whittling it down with a sharp knife, sanding it off with a belt sander or flapwheel grinder, or simply clamping the work piece to the edge of the bench and shaping the edge with a jack plane (Figure 1-18). It's really just a question of the tools you have at your disposal.

Once the rough shape of the two blades has been made, the edge can be shaped a little more with some 80-grit sandpaper on a sanding block (Figure 1-19).

At this point, the blades are about the right shape, as shown in Figure 1-20.

Unfortunately, they're still a little on the heavy side. This can be mitigated by cutting big chunks out of the middle of the blade with a hole saw, as shown in Figure 1-21. Just like in aircraft or ship construction, these interior holes (called *lightening holes*) will reduce the weight of the finished piece without compromising the strength.

FIGURE 1-19: Fine-tuning the blade shape with a sanding block

FIGURE 1-18: Cutting the edge of the blade with a jack plane

FIGURE 1-20: The rough-shaped blades

FIGURE 1-21: The hole saw quickly makes Swiss cheese out of the MDF axe blade.

No hole saw? Drilling a couple of starting holes will make it possible to cut big chunks out of the middle of the blade with a jigsaw.

The important thing to remember is to leave enough material along the outer edges for the next layers to adhere onto the finished piece. A ½" wide (10 mm) border should be more than enough.

Now that the two blade halves are shaped and the excess weight has been cut out of the middle, it's time to put them together. For this step, they need to be placed on top of a couple of blocks that will keep them both flat and parallel (Figure 1-22). Axe blades look funny if they're bent in the middle.

FIGURE 1-22: The blades propped up on some scraps of MDF so they'll stay straight

Now the cross pieces need to have holes cut into the middle in order for the handle to fit through. Then they can be slotted into the notches on the blades, as shown in Figure 1-23.

In order to make sure that the parts all stay aligned during the gluing stage, now's a good time to take advantage of some clamps (Figure 1-24).

The bar clamps will hold the whole assembly together while the glue is applied (Figure 1-25). Since MDF has a fairly porous surface, it can be

FIGURE 1-23: Keeping the handle in place makes sure that the holes will line up properly.

FIGURE 1-24: These are bar clamps, just one of the countless varieties of clamps that you will need more of.

held together with a cyanoacrylate adhesive such as Zap-a-Gap, Insta-Cure, or Super Glue. If those are too expensive, carpenter's wood glue will work just fine. It'll just take longer to cure and form a strong bond.

While the glue is probably going to be more than adequate to hold this whole thing together, it's always a good idea to be sure. In this case, it's a good idea to add a couple of screws to hold the bottom and top brackets onto the blade.

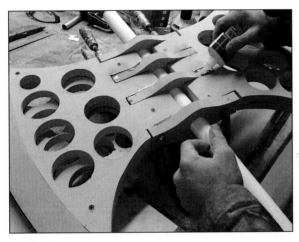

FIGURE 1-25: Applying glue to the seams

One of the biggest challenges when working with MDF is avoiding splitting the material when driving screws into it. Since it's basically just pressed wood garbage, it's really easy to wedge the bits of sawdust apart. This makes it a really good idea to pre-drill a hole everywhere a screw will need to be inserted. Start by drilling a small hole about the same size as the shaft of the screw (the skinny parts the threads stick out of), as you can see in Figure 1-26.

Then, since nobody wants to see screw heads sticking out of the ends of the giant battle axe, a bigger bit is used to make the top of the hole a bit bigger in order for the screw head to sink into the MDF (Figure 1-27).

Finally, drive a screw into that hole, secure in the knowledge that it won't do any more damage than absolutely necessary (Figure 1-28).

With the basic shape of the double-headed axe assembled, it's time to add a skin and make it pretty. Remove the handle and set it aside for now. Seriously. Resist the urge to do silly things with the as-yet-unfinished axe (Figure 1-29).

FIGURE 1-26: The pre-drilled hole will make it possible for the screw to effortlessly penetrate the MDF without splitting it apart.

FIGURE 1-27: Making a bigger hole for the screw head to hide in is called "countersinking."

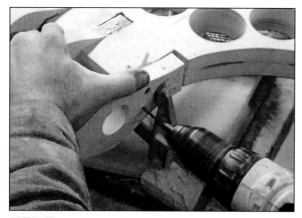

FIGURE 1-28: Driving a screw

Now it's time to cover the sides of the blade with a thin layer of sheet styrene. This can be purchased from plastic suppliers in various thicknesses. For smaller projects, you can use the plastic "for sale" signs available at your local hardware store. For even smaller projects, a variety of thicknesses of styrene sheet are available at your local hobby shop.

You're going to need a template and, in this case, a piece of butcher paper makes a good starting point. Simply hold it in place on top of the blade, and then use a crayon, pencil, or even a piece of chalk to do a *rubbing*. This means running the writing utensil back and forth across the piece. Everywhere there's a hard edge will become a darker impression on the surface, as shown in Figure 1-30.

Once the outline of the blade is clearly transferred to the paper, cut it out with a pair of scissors. This will give you the template you'll need for cutting the plastic.

FIGURE 1-29: Shenanigans. *Note: No silly little dogs were harmed in the taking of this photo. In fact, the only way to entice this one to run away was to throw a ball (visible in her mouth) and chase her as she went after it. These are the things we do for comedy.*

With the template cut out, trace it onto the sheet plastic, as shown in Figure 1-31.

There are a few good ways to cut the shape out of the plastic sheet. The first option: use a saw such as a jigsaw, scroll saw, or band saw. If you don't have access to any of these, run the tip of a sharp knife along the traced lines to lightly score the surface of the plastic. Once the lines are scored, you can either repeat this step a few more times until the blade cuts all the way through the sheet, or you can go back over it one more time to cut just a bit deeper into the sheet and then bend it until

it snaps along the scored line (Figure 1-32). Just be careful because this method will create very sharp edges.

Repeat this step to cut out four copies of the styrene shape (Figure 1-33).

FIGURE 1-30: Rubbing with a pencil

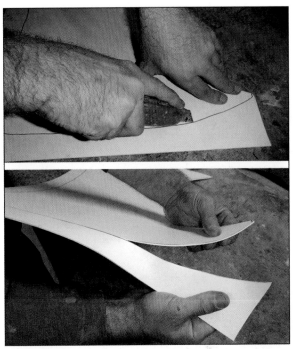

FIGURE 1-32: Score and snap—the quick and easy way to get shapes out of styrene

FIGURE 1-31: The shape traced onto the plastic sheet

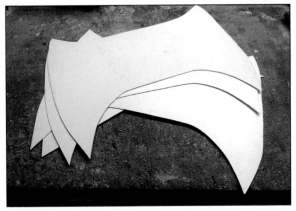

FIGURE 1-33: Four shall be the number of the counting, and the number of the counting shall be four.

Pick any of these four pieces and fit it into place on the MDF blade without gluing it in place. Here's where a handful of spring clamps will come in handy (Figure 1-34).

Dry-fitting the parts (assembling without glue) is a good idea. If it's not conforming to the curve where the handle will go through the blade, the sheet can be rolled by hand in order to pre-bend it so that it will lie in place better. Once the parts fit together neatly, you can apply glue to the MDF. The whole assembly can be clamped together securely while waiting for it to dry, as shown in Figure 1-35.

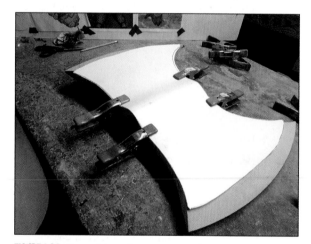

FIGURE 1-34: The world needs more spring clamps.

FIGURE 1-35: All of the available clamps. The more the better.

Repeat the fitting, gluing, and clamping for the other side as well. Once the glue is dry, the head of the axe is nice and smooth on the outside. The edges of the plastic may need some trimming to sit flush with the edges of the MDF, but that's a quick job for a knife. Since this is going to be a much more elaborately decorated piece with designs etched into the surface, however, it's going to need at least one more layer.

This last layer will end up making the higher areas bordering the etched recesses. This is a good time to cut out the recesses with a knife, or a jigsaw with a very fine blade.

When the pieces are cut, it's not uncommon for the edges to come out a bit rough. This can be readily solved with a piece of folded sandpaper. For the really tight corners, it's a good idea to get Mister Tongue Depressor involved (Figure 1-36).

Once all of the edges are cut out, it's simply a matter of adhering this final layer to the layer below. Start by dry-fitting to make sure everything looks right (Figure 1-37).

FIGURE 1-36: Mister Tongue Depressor (really just any small, flat piece of wood wrapped in sandpaper)

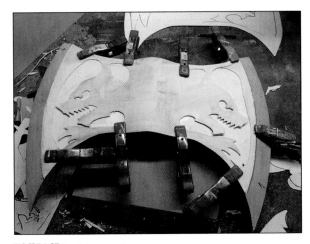

FIGURE 1-37: Looking good.

FIGURE 1-39: Scrap strips of wood can be used to clamp the layers together without scratching up the surface.

Before gluing on the final layer, now's a good time to attach the handle. This way it can be pinned in place with a screw. The screw hole will be covered up by the final layer of plastic sheet (Figure 1-38).

With the beginning of the handle glued in place, it's time to glue on the final layer of styrene. Since it may be a little tougher to get the parts to stick together, this is a good time to clamp thin strips of wood across the top of the parts to apply even pressure all the way across the work-in-progress, as shown in Figure 1-39.

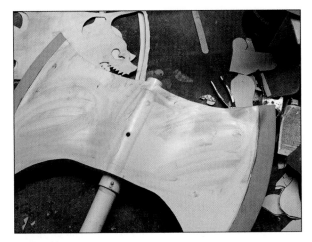

FIGURE 1-38: The handle installed with a set screw

🔫 Maker Note

While the layers of styrene can be held together with the same cyanoacrylate-type adhesive that was used to get the styrene to stick to the MDF, a better option is a "solvent cement," which you can get at hobby stores or the same supplier that sells the sheet styrene. This type of glue will chemically dissolve a bit of the styrene on the surface. Once the solvent evaporates, all that's left behind is the solid styrene, effectively welding the two pieces together. Even better, this water-thin glue can be applied using a syringe bottle or special applicator after the parts are clamped together (Figure 1-40). A drop applied at the edge will wick its way in between the parts by way of capillary action, meaning that the parts can be bonded after dry-fitting without disassembling them to add glue.

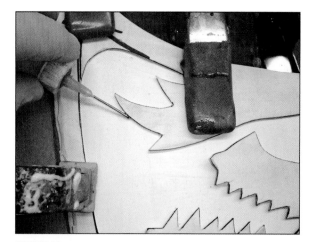

FIGURE 1-40: Using a syringe bottle to apply solvent cement to the seams

FIGURE 1-42: Smoothing the blade edge with a sanding block

Now that the head of the axe is assembled, there's still a bit of filler and sanding work to do. Since MDF is basically just sawdust held together with really weak glue, it'll be easy to nick or scratch the surface of the blade's edge. These edges can be saturated with more cyanoacrylate glue to make them durable (Figure 1-41).

As the glue soaks in and cures, the fuzzy sanded surface of the MDF will become rough and hard. Sand it smooth again with a sanding block (Figure 1-42).

Finally, it's a good idea to use some spot putty or body filler to fill in the edges of the blade and smooth over the seams between the various layers of MDF and styrene (Figure 1-43).

Now that the axe head is assembled and smoothed out, it's time to dress up the handle. This is a larger diameter piece of pipe with a PVC coupler glued to one end and a cap glued to the other end.

FIGURE 1-41: Soaking the edges of the blade with cyanoacrylate glue for strength

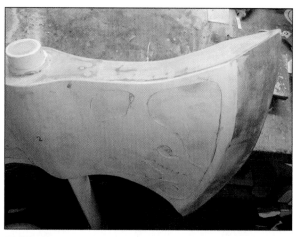

FIGURE 1-43: The seams along the edge need to be hidden somehow.

☞ Maker Note

Readily available PVC or ABS pipe fittings from the hardware store can be very helpful when building things that are cylindrical. The only problem is that a lot of these pieces will be easily recognized as pipe fittings. This can be solved by cutting them down in size, carving grooves into them, or grinding down the ends to make them less recognizable (Figure 1-44).

FIGURE 1-44: Boring pipe fittings (left) stop being recognizable (right) with just a bit of love from a miter saw and a sanding block.

In order to make the grip look less like an off-the-shelf piece of plastic pipe, carve a helical groove around it with a file and a sanding block. (Figure 1-45).

FIGURE 1-45: A groove carved into the handle to make it look less like pipe

Slide the grip assembly over the smaller handle and screw it into place (Figure 1-46).

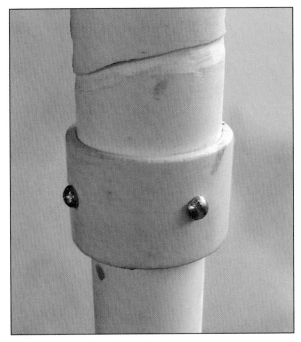

FIGURE 1-46: The handle screwed in place

☞ Maker Note

In order to keep everything from rattling around, the inner pipe can be shimmed in place by wrapping it with heavy paper or thin plastic scraps (called *shims*). Once the handle section is screwed in place, the gap between the inner pipe and the outer handle can be filled with a two-part epoxy putty available at most hardware stores. As an alternative, the gaps can be filled with auto body filler, casting resin, fiberglass resin, or a bunch of glue and sawdust. It's really just a question of getting the gaps filled with anything that will stay put once it cures.

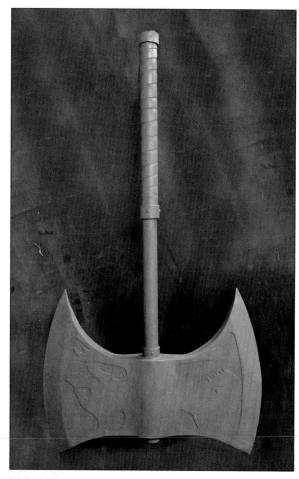

FIGURE 1-47: The completely assembled battle axe in gray primer

comic book character from the beginning of the chapter.

This thing is riddled with all kinds of little details and textures. To get it right, we're going to have to step up the game a bit.

Determining Scale

The first order of business with a project like this is to determine the exact right size of the finished piece. If there's a known dimension, such as the length overall, that's great, but if that information isn't available, it's time to do some educated guesswork.

If there's a projector available, this is a simple matter of projecting the profile image of the rifle on the wall. Then, put your hand where the handle is, as shown in Figure 1-49.

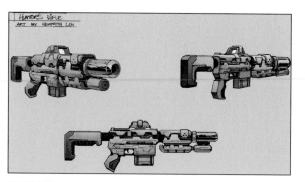

FIGURE 1-48: Not too shabby

Once the pieces are glued together, the whole thing needs a coat of primer and it's ready for paint (Figure 1-47).

A bit of paint can do wonderful things. We'll talk about painting in Part IV.

Hunter's Rifle

The great big battle axe was a pretty simple project. Just a few layers of flat material were needed to make some details stand out. For a bit more of a challenge, let's take a look at the rifle (Figure 1-48) carried by Hunter, the

FIGURE 1-49: A low-budget virtual test fit

Does it look too big? Too small? Zoom in or out on the image until it fits. Now take a ruler to the projected image and measure the overall length (Figure 1-50).

Now that you have the dimensions figured out, you just need to print out a life-size drawing to work with (Figure 1-51).

Since the whole thing has a bit of thickness to it, the main shape will be made out of a piece of ¾" MDF sheet. The barrels will be made separately out of different diameters of pipe. There will also be connections between the buttstock, main body, and fore grip area made out of sections of pipe to simulate the round parts.

For starters, trace the entire shape onto the ¾" MDF with a pencil. Then cut out the shape with a jigsaw, band saw, or coping saw and a lot of patience (Figure 1-52).

With the base shape cut out, now you've got a chance to run around the house with it making *pew pew* noises and frightening your pets and/or roommates. Since simply gluing pipe to the MDF wouldn't result in a very good bond, the front half of the main body will be notched out to allow a piece of pipe to be slipped into the end (Figure 1-53).

> ## 🔫 Maker Note
>
> It's a good idea to print a few extra copies. They're likely to get mangled.

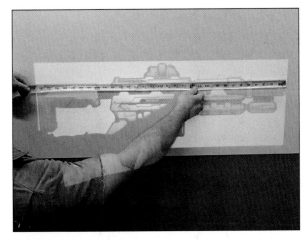

FIGURE 1-50: Measuring. With a tape measure.

FIGURE 1-51: The full-size paper drawing of the rifle

FIGURE 1-52: The jigsaw is the easy way.

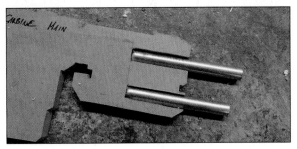

FIGURE 1-53: Notch with a piece of ½" metal electrical conduit fitted into it.

First Section: The Butt

Now that the first structural element has been figured out, it's time to dress the whole thing up. Since there are round sections connecting the buttstock to the main body, and the main body to the fore grip, it's going to make life a lot easier if we just cut those sections out of the base shape and build them as separate sections. Starting from the back and working forward, the easiest section of Hunter's rifle to make is going to be the back end, known as the *stock* or *butt* (Figure 1-54).

Now it's time to trace the next layer, cut it out, and glue it onto the base shape (Figure 1-55).

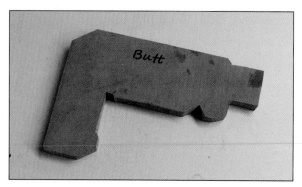

FIGURE 1-54: The butt. Hee hee.

FIGURE 1-56: Mister Tongue Depressor helps out with the finer things.

From there, each new layer adds a few more ridges and details (Figure 1-57).

The connection between the butt and the main body of the rifle is mostly round with a few raised details. This can be made by starting with a piece of PVC pipe (Figure 1-58).

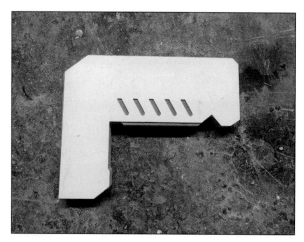

FIGURE 1-55: First new layer added

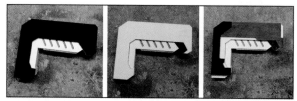

FIGURE 1-57: Fleshing out the details of the butt with more layers

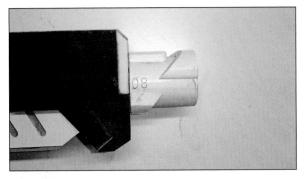

FIGURE 1-58: A short length of PVC pipe, dressed up with a few sections cut out of a pipe coupler

Because the whole thing was made of a stack of layered sheet stock, the edges will have a lot of visible seam lines. These will need to be filled in with auto body filler or epoxy putty and then sanded smooth (Figure 1-59).

Finally, in order to make it complete and ready for paint, the whole thing gets a coat of primer (Figure 1-60).

FIGURE 1-59: Smoothing over the edges to make the whole thing look like one piece

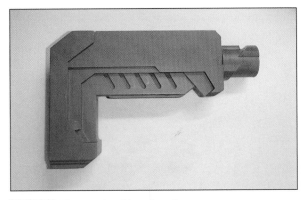

FIGURE 1-60: The completed butt in primer

The Main Body

The main body section in the middle is going to be a bit trickier. First let's take a look at the blocky, square pistol grip (Figure 1-61).

In order to make it a little more like something you'd want to hold with your hands, it's going to need to have a couple of the corners rounded off. But it's still going to have layers stacked on top of it and you don't want to take too much material off of the base shape. So, after cutting out the shape of the first raised layer, trace its outline onto the grip area (Figure 1-62).

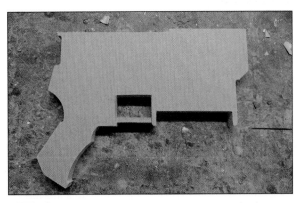

FIGURE 1-61: The base shape for the main body with the not-at-all comfortable grip

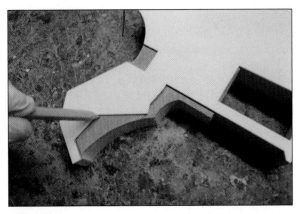

FIGURE 1-62: Marking the outline of the first raised layer of the grip

Then use a knife, file, rasp, or coarse sandpaper to round down the hard corners on the front edge of the grip (Figure 1-63).

Once the grip is rounded out as needed, it's time to start gluing on detail layers (Figure 1-64).

🔫 Maker Note

In order to make the ammo magazine removable, there is a notch in the main body forward of the grip. All of the layers after this will not have that notch. The end result will be a neat rectangular recess where the magazine will fit in. Nifty!

Planning for Lights and Electronics

Looking at the design references from the comic pages, there are supposed to be illuminated stripes running down the sides of the main body, as well as the forward section of the rifle. There are a handful of different ways to achieve this. The simplest would be to paint it with some top-of-the-line glowing paint. The most complicated would be to make the whole thing hollow, build in a couple of layers of transparent material, and then install an LED array inside that will backlit any exposed clear bits.

In this case, we'll be going with a pre-made electroluminescent wire system (Figure 1-65). Electroluminescent (EL) wire glows when a current is applied to it. It's not the brightest thing in the world, but it should create the just-right glow for this project. What's even better is that it can be purchased complete with a small battery-powered inverter that can be easily concealed inside props and costumes.

FIGURE 1-63: The front of the grip after whittling it round with a utility knife

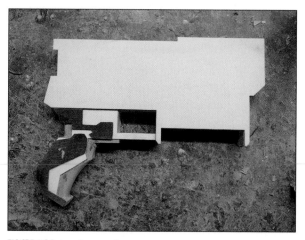

FIGURE 1-64: Stacking up layers

FIGURE 1-65: A neat little prepackaged EL wire system

In order to hold the EL wire secure in the finished piece, the next layer is going to need grooves cut to the same width as the wire (Figure 1-66).

FIGURE 1-66: Grooves for lights

🔫 Maker Note

EL wire can be cut to length, but adding lengths together can be a huge pain. With that in mind, these grooves were planned so that all of the lights are made from one continuous strand of wire. The strand will start from the top of the main body, snake its way along the grooves on one side, then go through a hole drilled to the other side, snake its way back up to the top, then go forward to fit into another set of grooves cut into the forward section. It'll all make sense soon.

Since the design doesn't show a single strip of light snaked all over each side of the rifle, the next layers have cutouts that will allow the light to only show through where it's supposed to be (Figure 1-67).

Since the EL wire won't be installed until after the whole thing is painted, these covers will not be glued in place until the very end of this project.

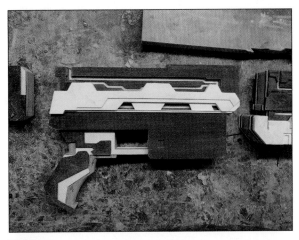

FIGURE 1-67: Covers designed to show light where it's needed

The Forward Body

The front section of the rifle is made in much the same way as the main body. It begins with those notches that were cut into it in order to fit a couple of lengths of metal conduit for strength (Figure 1-68).

Once the first few layers are glued on, the conduit is able to fit snugly into place, as shown in Figure 1-69.

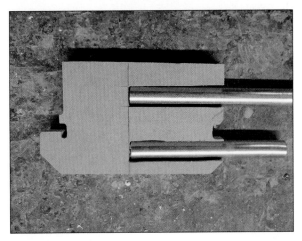

FIGURE 1-68: The base piece for the forward section of the body

FIGURE 1-69: The conduit held in place once the piece is skinned over with the first layers of plastic

Just like the main body, the layers are built up leaving grooves where the EL wire will fit in (Figure 1-70).

Finally, the bottom of the foregrip is shaped using files and a bit of auto body filler before being attached to the main body with another section of PVC pipe (Figure 1-71).

After a bit more sanding and touching up minor imperfections with spot putty, the whole thing is ready for primer (Figure 1-72).

The Barrel Assembly

The barrel section at the front of the rifle is made almost entirely out of miscellaneous PVC pipe fittings (Figure 1-73).

To make it harder for people to recognize the pieces as plumbing parts, it's a good idea to sand off any markings and mold lines (Figure 1-74).

Next, the pieces are cobbled together into a couple of basic sub-assemblies like those shown in Figure 1-75.

FIGURE 1-70: EL wire fitted into the grooves on the side of the forward section

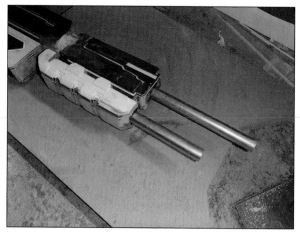

FIGURE 1-72: Smoothed out and ready for primer

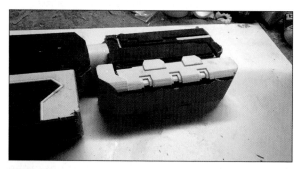

FIGURE 1-71: The nearly finished shape of the forward body attached to the front of the main body

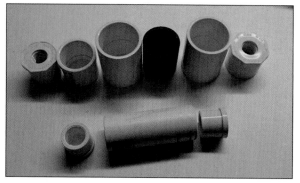

FIGURE 1-73: PVC bits are simple enough

FIGURE 1-74: Sanding off any incriminating markings on the fittings

FIGURE 1-75: Barrel parts coming together

🔫 Maker Note

When gluing together PVC parts, apply the glue to the inside of the female parts before assembly, as shown in Figure 1-76. If glue is applied to the outside of the male parts, it will get squeegeed off of the surface during assembly and end up being a visible flaw on the outside of the parts. There's nothing more embarrassing than squeegee goo on the outside of your parts.

FIGURE 1-76: Apply glue on the inside, not the outside

Once the basic shapes are put together, the front end of the barrel (aka the *muzzle*) is tapered with a sanding block to match the profile from the reference drawings (Figure 1-77).

After filling in the steps on the visible seam on the main barrel and sanding it smooth, some scraps of plastic are glued in place to form the bracket that holds the two pieces together (Figure 1-78).

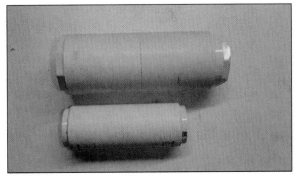

FIGURE 1-77: The basic shapes roughed out, including the tapered muzzle

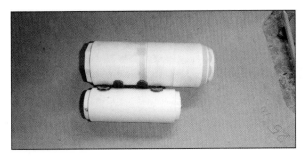

FIGURE 1-78: The finished barrel assembly, ready for primer

Upper Cover Pieces

The last thing to make is the cover that goes on top of the main body and the forward section of the rifle. These will be stacked up in layers just like the other pieces, with the outermost layers being tall enough to slide down over the main portions of the rifle, as shown in Figure 1-79.

FIGURE 1-79: The cover for the forward section fits snugly over the top of the rifle.

The main challenge in this case is to build the cover for the main body of the rifle with a cutout for the battery pack for the EL wire (Figure 1-83).

This cutout gets hidden under the sight assembly, which was also made out of layers of stacked plastic and wood (Figure 1-84).

Bending Sharp Corners in Sheet Plastic

The strap/clamp detail across the top of the sight was made out of a single strip of plastic that was bent to fit. In order to get a nice, sharp bend, the strip was placed on top of a toaster oven with the door propped open just a little bit, as shown in Figure 1-80.

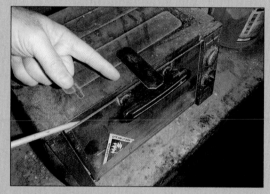

FIGURE 1-80: The gap in the door to the toaster oven concentrates released heat.

Once the area around the fold was good and hot, the strip was set in place on top of the sight, then the whole thing was set upside-down on the bench and the strip was bent upward until it fit tight against the side of the sight (Figure 1-81).

Next, the strip was placed back on top of the toaster oven to bend the other side. With that done, it became a nicely form-fitted bracket across the top of the sight (Figure 1-82).

FIGURE 1-81: Bending

FIGURE 1-82: Ta-da!

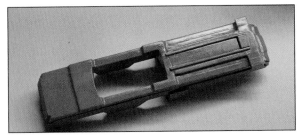

FIGURE 1-83: The cutout for the battery pack in the middle of the cover for the main body

FIGURE 1-85: Rifle parts all over the place

FIGURE 1-84: The sight fitted into place over the battery pack cutout on top of the rifle

FIGURE 1-86: Gluing the pieces together, with a straight edge in place for alignment

Final Assembly

At this point, the rifle amounts to a big pile of parts (Figure 1-85).

Final assembly started with gluing the barrels and buttstock in place, taking special care to ensure that everything was straight, as shown in Figure 1-86.

At this point, it's a good idea to paint everything before gluing in the EL wire and attaching the cover pieces. But there's no reason to resist the urge to put everything together and see how it'll look. In order to snake the EL wire from the main body to the forward section, drill a hole at an angle from the main body into the hollow pipe section (Figure 1-87).

FIGURE 1-87: Holes drilled to feed EL wire through

Once the wire is fed through, fit the covers into place and everything looks great (Figure 1-88).

Now it's time to take it apart and paint it. More on that in Part IV of this book.

FIGURE 1-88: *Pew-pew-pew!*

WORKING WITH PEPAKURA

Print a Science Fiction Helmet on Your Inkjet

A LOT OF AMAZING new technology is available to makers today. Technology websites and magazines are filled with countless articles about the latest developments in rapid prototyping and 3D printing.

What they don't say is that you don't actually *need* any of it for prop making and costuming.

For genuinely professional-looking results, all that's really needed is a computer with an ordinary printer; materials that are readily available at a local hardware store, hobby store, and office supply center; and some patience. Using a shareware program called Pepakura Designer—and techniques pioneered by the members of the 405th.com HALO costuming forum—we're going to walk through the creation of a wearable, science-fiction helmet prop (Figure 2-1).

"How?" you ask? Read on . . .

FIGURE 2-1: A helmet made out of PAPER

Pepakura Boot Camp

First, download and install a copy of Pepakura Designer from the Tamasoft website: tamasoft. co.jp/pepakura-en/.

The software doesn't *need* to be registered for it to work. But registration does unlock a few more functions, such as the ability to save work in progress. Besides, it's so inexpensive that it's worth paying the registration fee if only to encourage Tamasoft to keep making the program available.

Once the program is installed, the next thing you'll need is a 3D model to print out. If you're not a 3D modeler, never fear. It turns out you can find 3D models everywhere. The Tamasoft website has a gallery of models you can download for free, and there are countless prop and costume forums where members are willing to share models for use in Pepakura builds. If you have the appropriate conversion plug-ins, you can use models you design for yourself in freeware programs such as Blender or SketchUp as long as you can convert them to .OBJ or .STL format.

Once the program is installed, the next step is unfolding a model and printing out the pieces. The unfolding process can be tedious, but it's also a vital part of making a paper model that can be built practically. If you're starting with a model that's already been imported into Pepakura and unfolded, you're way ahead of the game. If you're starting with an .OBJ file that hasn't already been unfolded, there's a bit of work to be done. As an example, Figure 2-2 shows a simple 3D model of a cube with a notch in it.

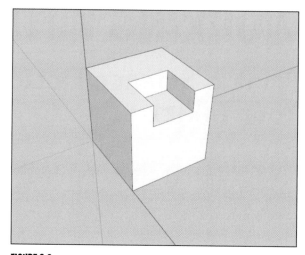

FIGURE 2-2: A 3D model of a cube with a notch in it

When the model is opened in Pepakura, the program will ask about flipping faces and front colors and a few other unnecessary things. Close out of this dialog box. Next, click the Unfold button at the top of the screen. A prompt will appear asking for the desired dimensions of the assembled model, as shown in Figure 2-3.

Once these values are entered, clicking OK will unfold the 3D model (shown in the program's left window) into the flat, oddball-looking thing in the right-hand window, as shown in Figure 2-4.

As you can see, Pepakura will try to make the model into as few pieces as possible so it can be assembled as quickly as possible. The problem with this is that computers will occasionally do perfectly logical things that make no sense at all to our imperfect, meat-filled heads. Fortunately, there are tools within the software that allow us to rearrange the seams (we'll explain those shortly). After a bit of dividing and joining edges, then repositioning the parts on the page, it starts to look like Figure 2-5.

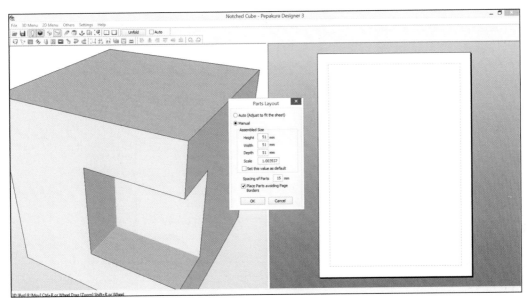

FIGURE 2-3: How big do you want it?

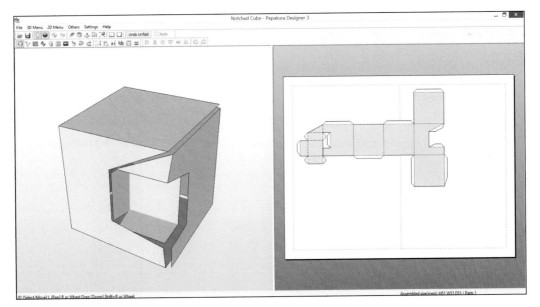

FIGURE 2-4: An oddball-looking thing (aka, a template)

Now that it's been converted to pieces that will all fit onto a single piece of paper, the tabs can be switched from one side to the other side of each of the seams as desired using the Flap tool, as shown in Figure 2-6.

Once the parts have been laid out and the tabs are set up as symmetrically as possible, it's time to print the object out and put it together.

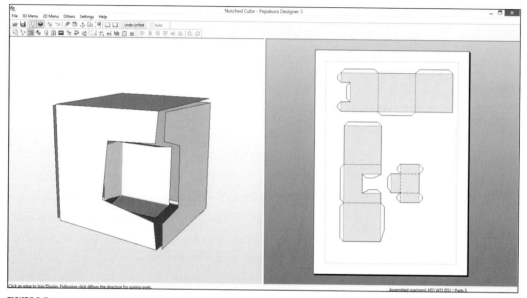

FIGURE 2-5: Something more sensible for meat-filled heads

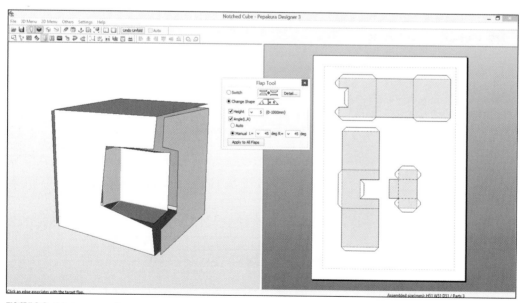

FIGURE 2-6: Editing some flaps

The parts should be printed on the thickest cardstock paper available (Figure 2-7). This way the parts will be less likely to sag and get distorted under their own weight.

With the printing done, it's time to cut the parts out of the sheet (see Figure 2-8). Sharp scissors or a hobby knife will come in handy. Cut along the solid lines only. The dotted lines are for later.

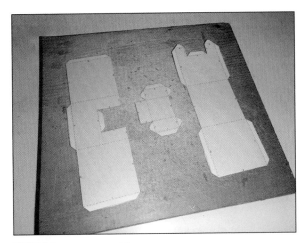

FIGURE 2-8: Pepakura model parts cut out

With the parts cut out along the solid lines, take a look at the dotted lines. These are the fold lines that indicate where the parts need to be creased. There are two types: the dotted lines indicating a *peak* fold, where the crease points upward, and the alternating dash-dot-dash lines indicating a *valley* fold, where the crease points downward. It's a good idea to score the folds to make it easier to get a nice, sharp crease. The easiest way to do this is to gently run the tip of your hobby knife blade along the fold lines, cutting slightly into the surface of the cardstock but not all the way through, as shown in Figure 2-9.

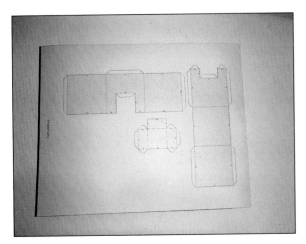

FIGURE 2-7: A simple Pepakura model printed on cardstock

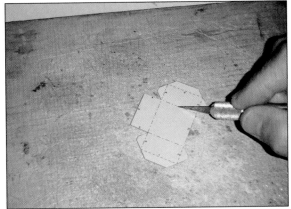

FIGURE 2-9: Cutting a little way into the cardstock

Now the part will fold easily along the shallow cuts (Figure 2-10).

FIGURE 2-10: A scored line makes for a clean, sharp fold.

It's time to start gluing the pieces together (Figure 2-11). When gluing the seams, the tabs should end up on the inside of the pieces, with the edge-identification numbers matched up evenly on opposite sides of the seams. Put a bit of glue on the flap, slide it under the adjacent piece so the edges are lined up evenly, and then hold it in place until the glue has dried.

As the pieces come together, there's no need to worry about exactly what angle they need to be positioned in. In almost every case, the geometry will work out so that there's only one correct way for the pieces to sit together without excessively warping the paper. The end result, shown in Figure 2-12, is a completely assembled piece that's the same shape as the original 3D model.

Chances are you're not going to find much reason to make a simple cube. If you use more complex

FIGURE 2-11: Gluing seams together occasionally means sticking your fingers together too.

FIGURE 2-12: A paper cube with a notch in it

models ripped from video games or downloaded from any of the many forums where users share digital models and Pepakura files, though, it's possible to make *much* more interesting things.

Putting Pepakura to Work

To go along with the science fiction rifle from the last chapter, I'm going to make the Hunter character's helmet (Figure 2-13).

Based on the concept art shown here, the folks at the 3D modeling service Do3D.com were able to design a digital model (Figure 2-14). You don't have to be an expert at 3D modeling yourself. Many 3D model websites offer custom services.

Once the 3D model is imported into Pepakura, it can be unfolded just like the notched cube. First, though, you'll need to determine a scale factor. Usually it's a simple matter of measuring the height, width, and depth of your head (or whichever body part the model is supposed to fit onto) and picking

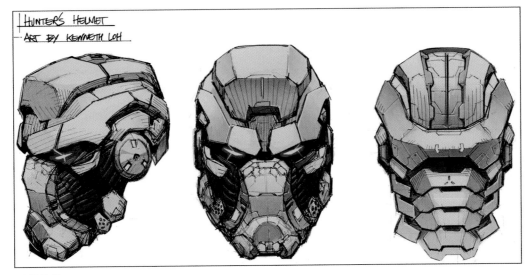

FIGURE 2-13: Hunter helmet concept art

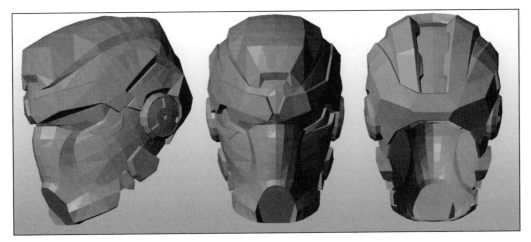

FIGURE 2-14: Hunter helmet 3D model—shiny!

dimensions that will allow for a bit of extra room inside. Some trial and error may be involved.

🔫 Maker Note

When choosing (or designing) your model, you have to weigh the complexity of the build against the amount of time and resources you'll have to spend making it nice and smooth. A high-polygon model will be harder to assemble in the paper stage. A low-polygon model will take more work when it comes time to smooth it out.

After plugging in the dimensions needed for the assembled model, click OK and the whole thing will be instantly unfolded to become the stuff of nightmares, as shown in Figure 2-15.

As expected, Pepakura begins by trying to make this significantly more-complex model into the fewest possible number of parts. To a computer, that might make sense, but to human makers (and those few extraterrestrial or artificial intelligences reading this book) it's going to be a major headache to make sense out of some of these pieces once they're printed onto cardstock (Figure 2-16).

It's time to start rearranging the seams on the paper model. If the model being used hasn't already been unfolded by one of the many prop-making saints who go around the Internet looking for models to unfold, this can be quite a chore. It's the same process used to rearrange the seams on the notched cube, though; it's just going to take longer this time around.

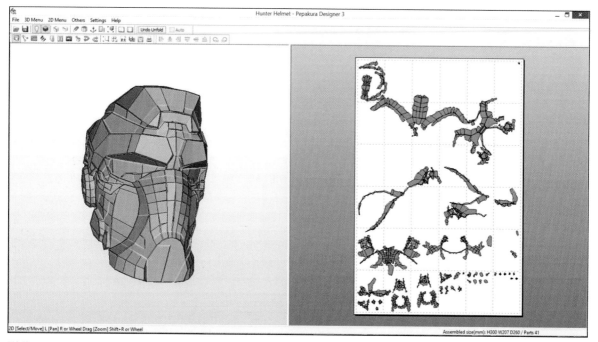

FIGURE 2-15: Eww! This is complicated!

To start with, select the Check Corresponding Face tool (hotkey Ctrl+K). Pick a readily identifiable part of the model in the left window and double-click to highlight it (Figure 2-17). This will show you the corresponding area of the unfolded template in the right window.

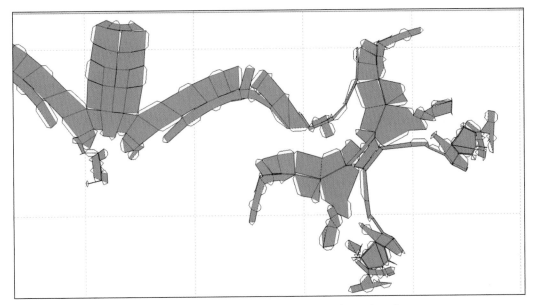

FIGURE 2-16: Evidence that the robot overlords want humans to get carpal tunnel syndrome

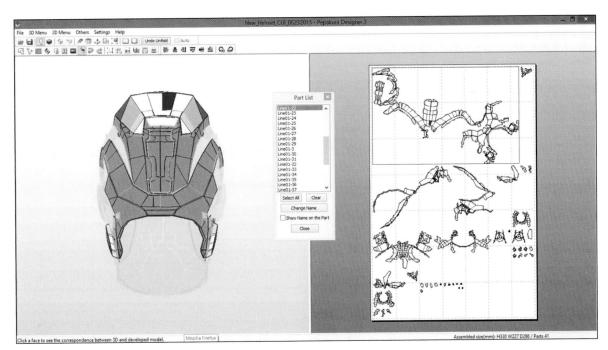

FIGURE 2-17: Finding the starting point

With a good starting point identified, it's time to start rearranging the seams to simplify the assembly of the model. This is primarily done using the Divide/Connect Faces tool (hotkey Ctrl+N). Figure 2-18 shows the crown of the helmet separated into a few parts that'll be easier to identify and assemble once they're printed on paper.

While most of those parts will be easy to cut out and glue together, the big piece in the middle has a lot of odd seams that will make it a bit tricky to get right. It's often easier to turn a big chunk like that into long strips, as shown in Figure 2-19.

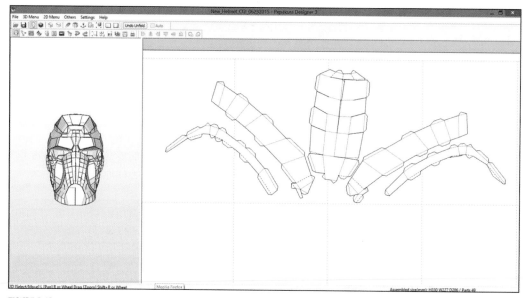

FIGURE 2-18: The crown parts

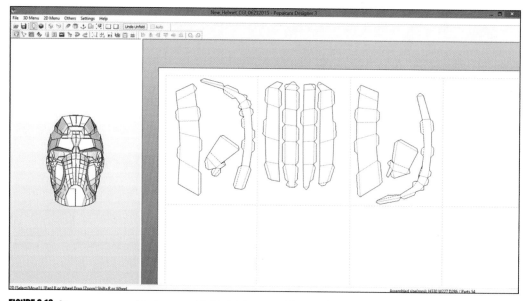

FIGURE 2-19: Seams rearranged to turn one big blob into several long strips

In the end, there's no easy way to get this job done quickly. Spending more time simplifying the build at this stage, however, will mean spending less time cursing at complicated pieces to cut and fold in the paper stage. In any case, after carefully rearranging the parts and putting the seams in places that make sense, the 2D window finally looks like something that can be built (see Figure 2-20).

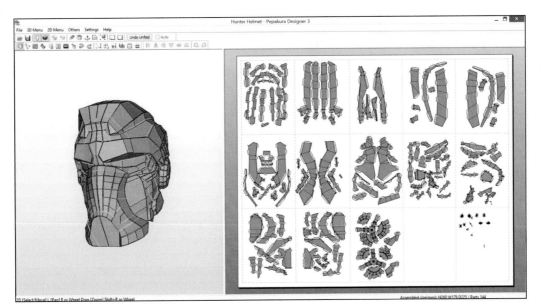

FIGURE 2-20: A paper model that human hands can build

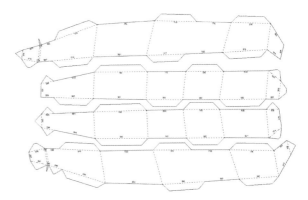

FIGURE 2-21: Tabs; like a zipper

The last little things to worry about are the last little things. Oftentimes, there will be a few tiny little pieces that, while they probably made sense on a digital model, are going to be nearly impossible to cut and fold in any useful way (Figure 2-22). You should drag all of these to a separate page in the bottom-right corner of the right window. Then don't bother printing out that page.

FIGURE 2-23: A clean, well-lit work area

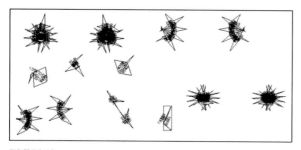

FIGURE 2-22: Tiny little parts nobody will ever cut out

With all of the parts laid out in a sensible manner on the 2D side of the window, it's time to print the whole thing out and get started. Here are a few important things to remember about printing the paper model:

- Under File ➤ Print and Paper Settings, be sure to check the Print Page Numbers box.

- If the printer chokes on a sheet of this unusually heavy paper, the sheet can be flipped over and reused.

- Once the pages are printed, keep them stacked and in order. It'll make life easier when searching for the next piece to cut out.

Now that the model is printed, get to building! Begin with a clean, well-lit work area like the one shown in Figure 2-23.

Then fill it with tools, as in Figure 2-24.

FIGURE 2-24: The Pepakura work area

Here are the building essentials you'll need to build your Pepakura model:

- The pages of the model printed out on heavy cardstock paper

- A shiny, sharp hobby knife

- Cyanoacrylate adhesive and accelerator

- An expendable cutting surface (tabletops and laps don't count)

- A computer with the model open in Pepakura to help keep track of work in progress

- Another device that can play a movie or music or something else to distract from the potential tedium of building the model

- Whatever kind of beverage will best facilitate the work

- Spare blades for the hobby knife and Band-Aids (optional, but probably wise)

For the novice, the next logical step would seem to be to cut out all of the pieces on the sheets. This would be madness. With all of the pieces cut out at once, the whole thing morphs into the world's most insane 3D jigsaw puzzle (see Figure 2-25) and also becomes a great reason to mock the noob who makes this mistake.

When building a Pep model, it's usually easier and wiser to hold off on cutting out the parts until you're actually ready to use them. Keep all of the printed sheets in order and use the Check Corresponding Face function to find the parts that are needed as assembly progresses (see Figure 2-26). With this tool selected, double-click on any facet in the model window and the corresponding facet will be highlighted in the 2D window.

FIGURE 2-25: The world's most insane 3D jigsaw puzzle

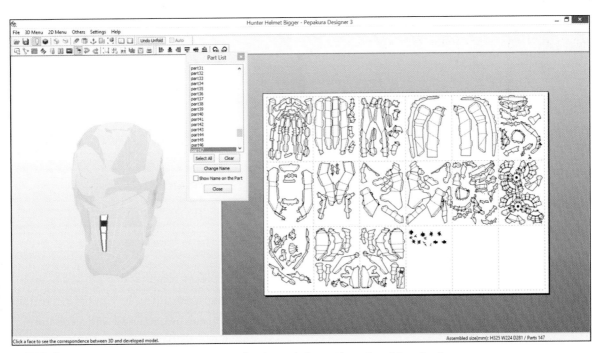

FIGURE 2-26: Using the Check Corresponding Face tool to instantly locate the next part to cut out

Picking a place to begin can be daunting. Usually it's a good idea to begin with the area as far as possible from the edges and work outward. It's a lot easier to add pieces to the outside of the work as opposed to trying to shoehorn them into the middle of a bunch of stuff that's already glued together. In this case, construction will begin with the nose area of the helmet. Start by cutting out a few pieces that will all end up glued together (as in Figure 2-27).

FIGURE 2-28: Gently scoring the fold lines with the tip of the knife

Once the pieces are scored along the fold lines, go ahead and pre-crease them (Figure 2-29).

FIGURE 2-27: Nose parts cut out

As each piece is cut out, use the knife to lightly score the fold lines so they'll crease easily (Figure 2-28). It's important to go along the center of the line and cut everything as precisely as possible.

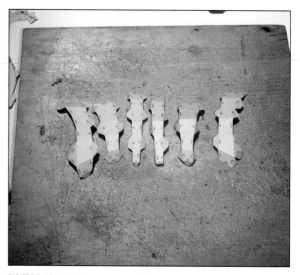

FIGURE 2-29: Folded pieces ready for assembly

🔫 Maker Note

When building a paper model like this, errors compound, so a few tiny misalignments along the way can add up to big weirdnesses later on. Neatly scoring along the middle of the fold lines will minimize weirdnesses.

Glue the pieces together one tab at a time using a cyanoacrylate adhesive. It cures in a matter of seconds and can be catalyzed with a separate accelerator spray if there isn't time for waiting. (There's never time for waiting.)

In just a few minutes, the nose is assembled (Figure 2-30).

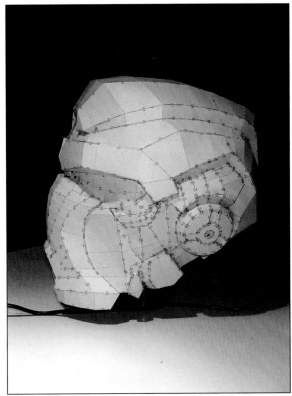

FIGURE 2-31: A flawless paper version of the 3D model

FIGURE 2-30: The assembled helmet schnoz

After a couple of evenings of cutting, gluing, cursing, and peeling fingertips off each other, the paper model is fully assembled (Figure 2-31).

Making It Hard

The fully assembled model may look pretty cool at this stage, but unless the character being built is a superhero whose one weakness is that their armor disintegrates in the rain, it's still going to need some work. The model also needs to be reinforced so it won't be crushed the moment someone looks at it the wrong way.

This is the time for fiberglass resin, which can be bought at your local hardware store. You'll need a disposable container to mix your resin in, as well as a disposable brush to spread it onto the surface. Muster your tools, take a deep breath, and read the label on the side of the fiberglass resin can. It's terrifying.

⚙ Warning

Polyester resins can be pretty nasty stuff. Before beginning this stage of the project, find an area outdoors to work or some place with plenty of ventilation to minimize exposure to fumes from the resin. Wear a respirator designed to filter out organic vapors to avoid making yourself stupid by huffing resin stink, and wear eye protection to keep from going blind if any stray splatter ends up near your eyeballs. Wear rubber gloves, and clothes you don't care about. They will be ruined during this process. Don't let any family pets eat this stuff either. It's bad for *them*, too.

Once again, it's time to start with a clear work surface like the one in my workshop, shown in Figure 2-32.

The model will be coated with liquid resin that will cure to form a sort of plastic. The problem is that, for a few minutes, the paper helmet will be wet. To prevent it from warping or sagging while soggy, it's a good idea to glue in a couple of cardboard braces to help keep everything aligned, as shown in Figure 2-33.

FIGURE 2-33: Cardboard solves many of the world's problems.

With the struts in place, cover your work area with something disposable (Figure 2-34). Wax paper works well, since you won't have to worry about it getting glued to the model.

Now that the work area is set up, it's time to gather the necessary tools (Figure 2-35). It's a good idea to have all of the tools handy and sitting on the bench. Once the fiberglass is mixed, there's only so much time to work, and it won't be a good idea to

FIGURE 2-32: A clean enough surface for fiberglass work

FIGURE 2-34: The workbench covered with a layer of cardboard. Wax paper would be even better.

frantically dig around looking for tools while wearing sticky, poison-coated gloves.

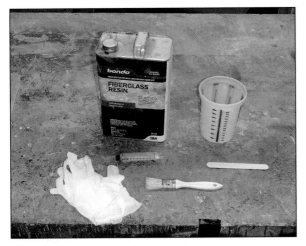

FIGURE 2-35: Tools for coating with fiberglass resin

Here's what you need for this stage of the project:

- Rubber gloves
- Disposable chip brush
- Mixing stick
- Graduated mixing cup
- Resin
- Catalyst

Mix a batch of resin in accordance with the manufacturer's instructions. For something the size of this helmet, six fluid ounces (about 175 mL) will be more than enough for one coat. In any case, the resin will begin to "gel" within about 15 minutes of mixing it, so don't mix more than can be used in that time. After adding the manufacturer-prescribed amount of hardening catalyst, use the stir stick to thoroughly blend it into the resin. Along the way, be sure to scrape the sides and bottom of the mixing cup to get the unmixed resin residue from the outside edges blended into everything else.

🔫 Maker Note

The mixing instructions for most fiberglass resin will include an optimal ambient temperature for the material to cure properly, as well as the ideal amount of catalyst to use. These values can be fudged a bit. Hotter days will cause the resin to harden faster, so it's a good idea to use slightly less catalyst. Conversely, colder days may require more catalyst to get the resin to cure in a timely manner. In fact, to a certain extent, slight (i.e., tiny, miniscule) adjustments to the amount of catalyst added can allow tailor-made cure times. This can be handy if you're in a hurry and need more time to apply a coat.

Use the chip brush to coat the cardstock with resin (Figure 2-36), taking care not to overdo it. At this stage, the resin doesn't need to soak all the way through the paper; it just needs to completely coat the outside.

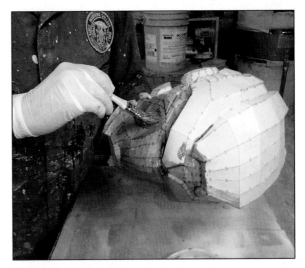

FIGURE 2-36: Making the paper shiny

It's also important to keep in mind that the resin shouldn't be dripping all over the place on the outside of the Pepakura model. A lot of work went into making the Pepakura model look as good as possible. There's no sense drooling a bunch of extra plastic goo all over it, only to grind it off again afterward. Done right, the whole thing should have just enough resin to lightly wet the entire exterior (Figure 2-37).

FIGURE 2-37: Shiny, resin-coated paper helmet

Once the resin has hardened, the helmet will be somewhat waterproof. It'll also be a bit stiffer, but it's not all that strong. Working with materials available at any hardware store, there are two especially popular options for strengthening it. The first one that a lot of people talk about is fiberglass layup. Fiberglass mat layup takes a bit of skill, a lot of time, and generates all kinds of mess, waste, and poisonous fumes. This is not a method for the beginner.

The other option that tends to generate better results for the novice is to coat the inside of the assembled Pepakura model with a blend of fiberglass resin and an auto-body filler commonly known by the brand name Bondo. Resin on its own is brittle. Bondo on its own is very thick and hard to slush around. Mixing the two of them makes a readily spreadable composite that's rock hard when it cures. This resin/Bondo mixture is often lovingly referred to as *Rondo*.

Determining the just-right mixing ratio between the two materials is largely a matter of personal preference. Lots of prop-making hobbyists will swear by their particular blending ratio. At any rate, there is plenty of room for experimentation. If a runny, watery mix is desired, use more resin. To make the mix more viscous, add more Bondo. In any case, once it's mixed, it'll start to harden before too long, so it's a good idea to work in small batches.

🔫 Maker Note

Bondo and fiberglass resin cure via an exothermic chemical reaction. This means that the mixture will generate heat as it cures. The bigger the batch, the greater the heat, and the faster it will cure. It might be tempting to mix one big batch to do all of the layering in one shot, but it'll likely just end up becoming a useless lump that'll need to be chiseled out of the mixing container.

Before mixing the Rondo, gather up your tools, shown in Figure 2-38.

FIGURE 2-38: Rondo coating tools

Here's what you'll need:

- Clean tool for scooping Bondo out of one-gallon cans
- Putty knives for mixing and spreading Bondo
- Chip brush for spreading Rondo
- Tongue depressors for mixing resin
- Clean, smooth surface for mixing Bondo
- Bondo
- Fiberglass resin
- One-quart graduated mixing cup

🔫 Maker Note

The gallon-sized cans of resin and Bondo are way too much for a project of this size, but it's always nice to have more on hand for the next project.

Once all of the tools are handy, it's time to put on some rubber gloves and mix up some goop. Here's how Rondo is made.

Step 1

Scoop some gray goop out of the Bondo bucket. This should *not* be done with the same tool that will be used to mix or spread the Bondo. That way, there's no chance of contaminating the rest of the can of Bondo with hardener and have it slowly turn into garbage. A helmet this size should only require a couple of blobs the size of golf balls. Or one blob the size of a baseball. Or two-thirds of a blob the size of a softball. It's not an exact amount, but balls are involved somehow.

Step 2

Add slightly less of the Bondo hardener than the instructions call for. The reason for skimping on the hardener is to allow a bit of extra mixing time before the material cures. Remember, you'll have to mix up a batch of fiberglass resin, too, so you don't want the Bondo to harden too quickly.

Step 3

Blend the hardener into the Bondo (Figure 2-39).

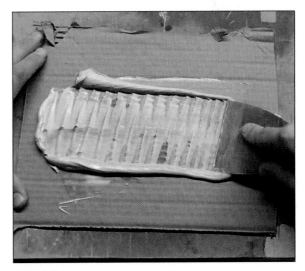

FIGURE 2-39: Blending hardener into Bondo

Secrets of Bondo

If your workshop is as well equipped as mine, you have a retired auto mechanic father who stops by whenever he gets bored with his antique car restoration projects to tell you what you're doing wrong. If you don't have such help, here's what Dad usually tells me when I'm mixing Bondo:

- Keep everything neat and clean.

- Let the Bondo touch only one side of the putty knife so you can control where it goes. If it ends up on the backside of the putty knife, it'll trail little ribbons along the top of your work and it'll be impossible to spread it smoothly.

- While mixing, periodically scrape the Bondo off the working side of the putty knife so it can be blended back into the rest of the material.

- Scrape the Bondo off the mixing surface from time to time, as well, to make sure the stuff on the bottom gets mixed in, too.

- Continue mixing and scraping and folding the Bondo back into itself until it's all one uniform color.

- Make smaller batches. When you sand it all off, you won't waste as much time and money filling the workspace with pink dust.

- Get a job.

That last one is apropos of nothing. It just comes up a lot when my Dad visits.

Step 4

Once the Bondo and hardener are thoroughly mixed to an even, homogeneous color, mix up a batch of fiberglass resin (in accordance with the manufacturer's instructions) that's about the same volume as the blob of Bondo. Then scoop the Bondo into the resin (Figure 2-40).

Step 5

Using a tongue depressor or other flat-sided mixing tool, mix the resin and Bondo together until it has an even color and consistency, as shown in Figure 2-41.

FIGURE 2-40: Blob of Bondo blobbed into mixing cup of resin to make a superblob

FIGURE 2-41: The world's nastiest smoothie

When it's mixed thoroughly, it becomes a blended goop with a viscosity somewhere between milk and mayonnaise, depending on the mixing ratio. More resin equals thinner goop. More Bondo equals thicker goop.

This goop is RONDO!

Step 6

Pour the Rondo into the Pepakura helmet, as shown in Figure 2-42.

Step 7

Tip the work piece from side to side and front to back so the Rondo slushes around and coats everything on the inside. The goop will eventually cure to a solid, almost rock-hard mass. The object is to keep moving the helmet around in order to get the goop evenly spread over the interior of the paper. Otherwise, the Rondo will drool its way down to the bottom of the helmet as shown in Figure 2-43 and solidify into a big, heavy, thick area that will never sit right and make the helmet lopsided and unbalanced. This can be prevented

by continually keeping the whole thing in motion while letting the Rondo slosh and splatter around the interior until it gels and hardens into a shell of uniform thickness.

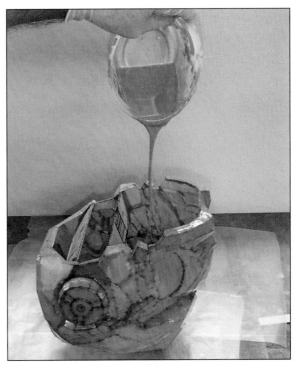

FIGURE 2-42: The world's nastiest smoothie poured into a paper hat

FIGURE 2-43: Keep moving the whole thing in order to keep goop from pooling in one place.

During this stage of the project, it's absolutely guaranteed that at least a little bit of liquid Rondo will drip out somewhere. Be sure to cover the floor with the finest antique Persian rug available, let it soak up any spills, then leave it where it is until the Rondo has cured and it glues the rug to the underlying floor. Actually, that's a bad idea. The better idea is to lay down a bed of newspaper or wax paper over the work area so it can all be rolled up and disposed of when the project is done.

Step 8

Mix up another batch of Rondo and repeat the coating process as many times as needed to make the helmet nice and strong. After enough iterations, it'll be pretty sturdy (Figure 2-44). How sturdy? That depends on how many coats you layer on. I may have overdone it with this one.

There is no trickery involved in this picture. That's all 175 pounds of me plus steel-toed boots, heavy coveralls, and pockets likely filled with sanding dust and used rubber gloves standing on top of this paper helmet. I may have overdone it with the reinforcement, but it gives you an idea of what's possible.

Making It Smooth

So all it takes is some paper and hardware store materials to make a nice, strong helmet. It's not safe to wear for motorcycling, or hockey, or getting shot out of a cannon, but it should be more than adequate to hold up to the rigors of costuming. The next step is to smooth the outside so it stops looking like a multifaceted 3D model and starts looking sleek. This, too, is a job for Bondo body filler.

Remember all of the guidelines mentioned above about mixing and working with Bondo? They still apply. All of the required "body shop" tools are shown in Figure 2-45.

If you're paying attention, you'll notice that these are basically the same tools that you used in the Rondo phase, minus the fiberglass resin, tongue depressors, and mixing cup. The only new tools are the files, sanding blocks, and sandpaper that will be used to shape the cured Bondo.

With the tools gathered up, it's time to set up the workspace. This step is going to take a lot of time and generate a lot of dust, so it should be done somewhere that can deal with that kind of

FIGURE 2-44: Standing on top of the Rondo-reinforced helmet

FIGURE 2-45: The essential "body shop" tools

mess. Since the aim is to take the faceted, digitally generated paper model and smooth out all the surfaces to make them look like the original character design, it's also a good idea to have all the available reference images close at hand. (Your workspace should look something like what you see in Figure 2-46).

FIGURE 2-46: Find a well-lit work area where nobody will mind the dust. All of the reference images should be readily available.

Now that everything's ready to go, it's time to get to work. I usually start by using a wood rasp to grind off the high corners of any area that's going to need to become curved. Then the first layer of body filler goes on. If the initial 3D model was already fairly smooth, it shouldn't need very much filler to fine-tune the shape. If it started as a low-polygon model, it'll need more filler to round out the facets. In either case, resist the urge to pile on a lot of material. It'll just need to be ground down as soon as it cures. Instead, it's easier (and less messy) to build up several small layers and cut down on the amount of sanding time and material waste along the way.

With the first coat of Bondo on the Hunter helmet, it looks like Figure 2-47. Note that there's not

just a huge glob piled on willy-nilly that would then need to be carved back down. As the Bondo cures, it will become progressively firmer, allowing the user to do some rough shaping with the putty knife. At this stage it's a good idea to shape it as close as possible to the finished form, but don't fret if it's not perfect. There will be plenty of time for perfect later.

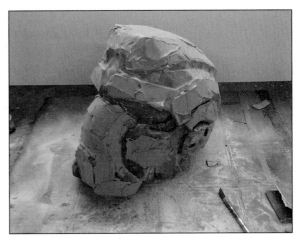

FIGURE 2-47: The first coat of Bondo

After about 20 minutes, the filler will harden to the point where it can't be smooshed around with the putty knife any more. After another 20 minutes, it can be carved and sanded to knock down any high spots or rough patches that may have come up during the smooshing. The best tool to start with is called a Surform shaver; it's the little cheese-grater-looking widget with a bright yellow handle shown in Figure 2-46. They come in a variety of shapes and sizes, but the little rounded one will remove a lot of material in a hurry at this stage (Figure 2-48).

This is the tedious stage of the project. From here the next step is to continue filling in the unwanted low spots and sanding down the unwanted high spots. There will be curves that need to be made straighter or flat, and faceted areas that need to be evened out to create compound curved areas. The general progression of sanding and filling with the Hunter helmet is shown in Figure 2-49. Get comfortable and settle in for the long haul. Unless you're some kind of sanding savant, this is the stage that will consume the largest number of hours.

FIGURE 2-48: Smooth, but not quite smooth enough

Somewhere in the seemingly endless loop of smoothing and shaping and filling and sanding, it'll eventually reach the "zit stage." This is the point at which continuing to mess with it will only make it worse. It's important to recognize this stage when it is reached; otherwise, it's easy to get lost in a hellish, dust-filled downward spiral that will take over your life like some chalky, time-sucking leech.

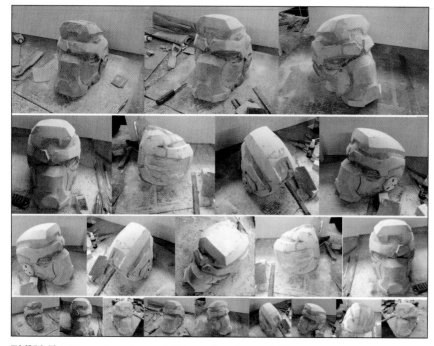

FIGURE 2-49: Filling and sanding and filling and sanding and filling and sanding and filling and sanding . . .

It may even require the timely intervention of friends and family and possibly the occasional mental health professional who can, after a lengthy and emotional intervention, somehow convince you that it looks "pretty good."

At this point (or much sooner probably), it's a good time to spray on a coat of primer to identify areas that need a little more attention (Figure 2-50).

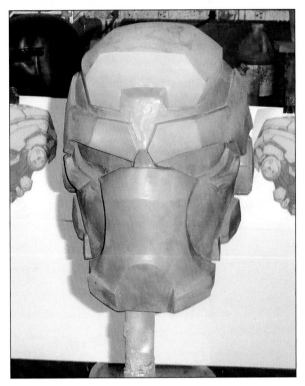

FIGURE 2-50. A coat of gray primer, the great revealer

While things may have looked "pretty good" when they were all sorts of different colors and covered with dust, a coat of primer will take away any deceptions caused by the different colors and show all of the remaining flaws and problem areas. Now it's time once again to set off on the long cycle of filling, sanding, filling, sanding, and filling, and sanding.

During the course of the sanding, filling, and priming process, there will likely be places where the original paper bits become exposed. When these areas are sprayed with primer, they'll likely end up soaking in a bunch of the primer and looking fuzzy and weird. Not to worry. Once the primer dries, sand these areas lightly with some fine sandpaper (220-grit should be good) and then spray on another coat of primer. A couple of rounds like this and it should be good to go.

When all of the little problems are finally fixed, prime it again. If it's smooth and straight and there's no more visible fuzzy paper showing, it's time to move on to detailing.

Detailing

Now that the basic shapes are smoothed out and curved or flattened as needed, it's time to add details. There are usually at least a few bits and pieces that need to be added to the physical model that weren't present on the digital model. This is especially true of a lot of the models that were originally designed to render characters or props in video games. In most cases, developers would rather not waste rendering power on tiny seams and details when they can just be painted onto the skin that'll be visible in the game.

Small seam lines and grooves can be carved in with a triangle file or, just as well, with a loose jigsaw blade. Use a straight edge and a hobby knife to lightly scribe a mark where the grooves will be. Then go back over the scribed lines with a small file or saw blade to make them deeper and wider as needed (Figure 2-51).

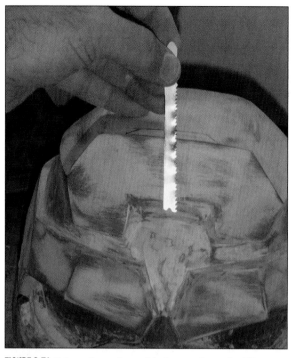

FIGURE 2-51: Using a loose jigsaw blade to make scribed lines groovy

Not all of the details will be this simple, though. If there's a raised edge that wasn't present on the model, more Bondo can be used to build up an area. On the Hunter helmet, the cheek areas are covered with a bunch of ridges that were not present in the model. In order to ensure that they all have the same cross-section shape, a *contour gauge* was carved out of the end of a mixing stick (Figure 2-52).

FIGURE 2-52: A contour gauge made by whittling the end of a mixing stick

Then the cheek area was covered in Bondo and shaped by dragging the contour gauge through it. This leaves a series of matching ridges (Figure 2-53).

FIGURE 2-53: Building up rows of ridges. Like a boss.

🔫 Maker Tip

If the design calls for bolts or screw heads to be visible, the easiest way to simulate them is to pick them up at the local hardware store and just drill holes to mount them onto the assembled piece.

These final touches can be added after the bulk of the shaping and sanding has been done, but there will still be occasional flaws to fix. Ideally, the bigger problem areas have been fixed with Bondo. Most of the flaws at this point should be fixable with glazing and spot putty, another automotive product available at the local hardware or auto parts store. It's designed to fill in pinholes and tiny scratches before you spray on your paint. It dries on contact with air and will shrink slightly as it dries. For this reason (and the fact that it costs money) spot putty should only be used to fill in the tiniest of tiny holes (Figure 2-54).

FIGURE 2-55: DONE!

FIGURE 2-54: A few tiny holes filled with spot putty

Once the holes are filled in and the putty has dried, sand it smooth with 220-grit sandpaper. Then spray the whole thing with yet another coat of primer (Figure 2-55).

If the plan is to just make one helmet to wear around and look awesome, skip ahead to Part IV, "Painting and Weathering," to learn how to finish the whole thing. If you have reason to make more than one, Chapter 8, "Molds for Rotocasting," explains how to make another copy of the assembled helmet without going through this entire process again.

EVA FOAM

Turning Floor Mats into Armor

IN THE LAST CHAPTER, you saw the amazing things you can do with paper and Pepakura. In this chapter you're going to learn how to turn humble garage floor mats into a convincing suit of armor. This is the part where you're probably thinking something like, "I don't want to make a suit of armor out of foam! I want to make it out of hammered steel or a space-age carbon fiber and Kevlar composite with ceramic backer plates so it'll be rock hard and stop bullets. After all, it's supposed to be armor."

If that's what you're thinking, you haven't given it nearly enough thought.

Making your costume armor out of hard materials adds several orders of magnitude to the level of fabrication difficulty. It also means using a whole bunch of specialized tools and making a much bigger mess in the process. Also, the unyielding parts bumping and grinding against each other as you walk will tear up the paintjob faster than you can say, "Ow, that pinches!" Finally, while it might sound cool to be able to say something like, "My futuristic galactic space marine armor is made out of actual steel," you're going to suffer some ironclad embarrassment when you pass out from exhaustion after lugging all the extra weight around in the sun at some event.

It's better to fake it. Done right, it'll be just as awesome to look at without all the extra pain and suffering, pinching, bleeding, heat exhaustion, paramedics, and hospital bills.

So how do we make rubber mats into amazingly convincing suits of armor? Well, it's important to learn how to crawl before you can sprint down the runway on the way to a jetpack-clad takeoff. That's why we'll start with . . .

Sourcing: Where to Get Foam

Asking around for "EVA foam" at local stores will get a lot of confused responses. Instead, ask for it based on what it's used for. The most common version of thicker EVA foam will be anti-fatigue floor mats, which can be found at most hardware stores and auto parts stores. They'll be sold either in large rolls or in interlocking tiles ranging in thickness from 3/8" to 5/8" (about 1 cm). They're available in a variety of colors, but most commonly are they'll be some shade of gray. It will usually have one smooth side and a textured side designed to keep people from slipping on it.

Most of the time, a store will only have one or two types of foam floor mats. If the plan is to incorporate any of the textures into the finished piece, it's a good idea to buy foam from more than one store to give you a variety of textures.

The other common version is often called *craft foam* or *Foamies*, and can be bought in most craft stores. It's usually around 1/8" (3 mm) thick and comes in all sorts of different colors.

Just a little bit of shopping around and you'll find you have all sorts of choices in thickness, color, and texture available to you (Figure 3-1).

FIGURE 3-1: There's a whole world of EVA foam just waiting to be discovered.

Now that you have yourself some foam, here are the other tools and materials you need:

- Cutting tools (e.g., box cutter, utility knife, a hot knife, and/or a scroll saw)
- Contact cement
- Paintable caulking
- Butcher paper
- Duct tape
- Aluminum foil
- Permanent marker
- Straight edge
- Sandpaper
- Heat gun
- Blowtorch
- Plasti Dip or Mod Podge
- Rotary tool (such as a Dremel)
- (Optional) Computer and printer
- (Optional) Blowtorch

Once you have your tools and materials, let the foam armorsmithing begin!

The Basics: Making Foam Obey Your Every Whim

Before we dive into the how-to of making foam armor, let's look at some of the mandatory methodologies to master this material.

Cutting the Foam

Unless you plan to dress up as "Jigsaw, the Madman in the Interlocking Diamond-Plate Armor," chances are good that the foam isn't sold in the right shape. Instead, it's time to do some cutting.

In most cases, this just calls for a very sharp knife, such as a box cutter or a utility knife with a fresh, sharp blade. To make a straight cut, start by pressing the tip of the blade all the way through the foam, then drawing it along the line to be cut at the shallowest angle possible (Figure 3-2).

For curved cuts, it's pretty much the same operation, but the blade is held at a steeper angle (Figure 3-3).

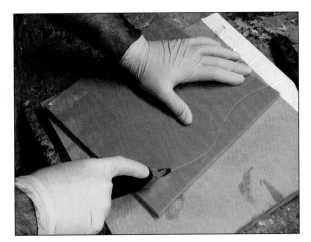

FIGURE 3-3: Cutting a curve

Maker Note

When using a knife, be sure to keep the blade as sharp as possible. That said, eventually there will be occasions where the foam edges will tear or skip slightly while cutting. These edges can be sanded smooth with a sanding block or with a sanding drum on the rotary tool in order to get rid of the resulting roughness (Figure 3-4).

A hot knife (a type of cutting tool with an electrically heated blade) can be a good idea, as well. Just remember to cut fast enough to avoid burning the foam and slow enough to avoid tearing it. The hot knife has the advantage of heat sealing the

FIGURE 3-2: Making a straight cut in foam

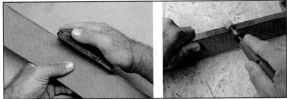

FIGURE 3-4: Smoothing the edges with a sanding block (left) or the rotary sanding drum (right)

edges while it cuts, eliminating much of the porosity of the cut surface.

A band saw or scroll saw (Figure 3-5) can speed things up too. At the very least it'll make things a bit easier on the fingers.

FIGURE 3-5: The scroll saw makes quick work of cutting foam.

Joining Seams

Once the parts are cut out, they'll need to be glued together. There are a few preferred methods for doing so. The first, and most talked about, is contact cement. This is a readily available, air-dry adhesive that can be bought at just about any hardware store.

Before using contact cement, it's a good idea to heat-seal the edges of the foam. This means using a blowtorch or heat gun to quickly melt the surface of the foam and eliminate the tiny pinholes that were created when cutting through the cellular structure of the foam. If the edges aren't heat-sealed, they'll soak up a lot of the glue and it'll take multiple coats (and a lot more time) in order to get a good bond.

As always, READ THE INSTRUCTIONS on the glue before using it. Here's the short version: use a brush to coat the two surfaces to be joined with contact cement, allow them to dry until tacky, then push them together and allow them to dry completely. This will result in a strong, reliable bond. The only real trick is making sure that the outer surfaces of the two parts meet up evenly in order to keep the seam nice and neat (Figure 3-6). The main drawback to contact cement is that it's a bit slow to dry.

🔫 Maker Note

In most cases, the contact cement will be soaked into the surface of the foam. This just means that a second (or even third) coat will need to be applied and allowed to dry until tacky before joining the parts together. A tiny bit of penetration like this will make for a very strong bond. It's a good thing.

The second option for sticking the foam together is to use a hot glue gun. This is a faster method, but has a few major drawbacks. The main one is that the glue will melt at high temperatures. This isn't usually a problem, but out in the hot sun in a thick suit made of foam, the temperature inside can get pretty warm rather quickly. Eventually it may be warm enough to soften the glue just a little bit. When the glue gets softer, it gets weaker. Then the armor will start to come apart a little bit. Then it will look a little less awesome as it begins to slowly disintegrate. Again, it's not usually a problem, but it's something to be aware of.

The second problem with hot glue is that some brands are water soluble to some extent. It may

FIGURE 3-6: Joining two pieces with contact cement

seem like this wouldn't be a problem unless the armor were for some sort of submarine scuba soldier, but water just happens to be one of the main ingredients in sweat. So once again, on that hot day . . .

The third (and biggest) problem with hot glue is that it can have a lot of bulk to it. While most adhesives are just a thin coating added onto the surfaces to be joined, hot glue is a blob of melted goo that gets put in between. When the parts are pressed together, all of that goo has to go somewhere, so be careful when squeezing parts together so it doesn't leave an unsightly blob somewhere that'll be visible. It's not the end of the world, but it's something to take into account when using it to put parts together.

Another option for adhesive is the same cyanoacrylate adhesive that worked so well when building Pepakura models. Often called *super glue* and available at any hardware store, hobby shop, or craft store, it cures quickly and the cure time can be made much faster with the use of an accelerator

usually referred to as *kicker*. While it's not a terrible option, its fast cure time can be a problem if the parts aren't positioned perfectly on the first try, and it tends to be hard and brittle, which will become a problem if the parts will need to flex and bend a lot while being worn.

Whichever type of adhesive is chosen to construct the suit, it's always a good idea to reinforce the seams with a little bit of fabric backing in order to help take the strain. We'll cover this in more detail later.

Seams, Joints, or Whatever You Want to Call the Connections

The easiest way to stick two pieces of foam together is a *butt* joint, where the two pieces butt up against each other, as shown in Figure 3-7.

If the seam is supposed to correspond to a hard corner on the surface of the piece being made, it calls for a *miter* joint. To make a miter joint, start by cutting the edges of the parts at an angle, then joining them together, as shown in Figure 3-8.

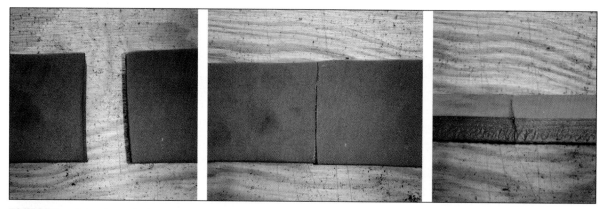

FIGURE 3-7: Butt joint. Stop giggling.

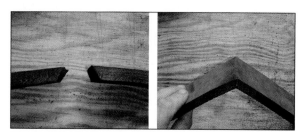

FIGURE 3-8: Viewed from end on, the angle-cut (or *mitered*) pieces join together to create a corner.

Finally, a *lap* joint is formed when two pieces are put together so they overlap instead of joining at the edges (Figure 3-9).

Once the seams are joined, there will probably be a visible edge where the parts were glued together. While careful work can sometimes make them hard to spot, it's often a good idea to do a little bit of filling and smoothing work in order to help hide the seams. While shopping for foam and sandpaper, pick up a small tube of caulking. Not just any caulking, though; make sure to buy flexible, paintable, sandable caulking. This can be used to fill in the tiny surface gap where the pieces come together.

To hide the seam, start by squeezing a bead of caulking along the length of the seam. Then, dip your fingertip into a cup of clean water. Use the wet fingertip to get the caulking as smooth as possible, then set the piece aside and allow the caulking to dry.

After the caulking has dried, sand the area lightly with medium- to fine-grit sandpaper to blend it in even further.

FIGURE 3-9: A lap joint is a good idea for pieces that have visible steps or variations in thickness.

Corners, Bends, and Compound Curves

Suppose there's a place where the foam needs to have a tight bend with a small radius, but there's no need for a seam along that edge. The foam can be made to bend tightly by cutting out a groove in the back side and then bonding it together afterward. Start by bending the foam sheet in the opposite direction of the corner that will be needed, then cut off some of the inside with a knife, as shown in Figure 3-10.

> ### 🔫 Maker Note
>
> Instead of cutting it with a knife, the inside can just as easily be ground out with sandpaper or a rotary tool.

After removing a slice of foam from what will be the inside of the rounded corner, coat the gap with adhesive and fold the foam back on itself. Once it cures, this will leave a rounded corner on the outside of the foam (Figure 3-11).

FIGURE 3-10: Cutting the inside out of what will become a rounded corner

FIGURE 3-11: After gluing the inside corner, the outside is left with a nicely rounded edge.

If the final shape needs to be a bigger, longer curve instead of a tight, rounded corner, it's time to turn up the heat. Like any plastic or rubber material, EVA foam can be melted at relatively low temperatures. At temperatures slightly below melting, it can be deformed and reshaped to a limited extent and then, after it cools, it will retain this new shape.

> ### 🔫 Maker Note
>
> The foam melts and deforms at *relatively* low temperatures compared to glass or steel. The heat gun will still generate enough heat to peel the paint off of the walls at close range, so exercise caution. Don't leave it turned on and unattended, and remember that after it's turned off, it'll still be hot for a while, so be careful where you set it down when you're done with it. You've been warned. If you burn down your house and destroy everything you love in the process, it's your own fault.

To reshape the foam by hand, start by heating the foam with a heat gun. The idea is to increase the temperature of the piece without burning a hole in the surface along the way, so be sure to move the heat gun around while applying heat to the foam, and don't forget to heat the sheet on both sides. Holding still for any length of time will potentially melt or burn a hole in the foam.

It really doesn't take all that much heat before the foam will become warm enough to bend permanently, as shown in Figure 3-12.

FIGURE 3-12: After you warm the strip with the heat gun and bend it by hand, it will keep its new shape when it cools.

Using carefully applied heat, and bare (or gloved) hands, it's possible to get the foam to conform to almost any shape (Figure 3-13).

Compound curves can also be achieved in much the same way. Simply heat the surface of the foam and stretch it by hand or force it over whatever object is handy until it has the necessary shape (Figure 3-14).

FIGURE 3-13: No idea why anyone would want a piece of foam shaped like this, but here it is.

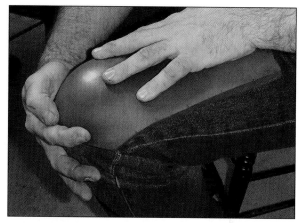

FIGURE 3-14: Pushing a bump into the heated foam by using a knee as a form

Patterning

There are basically three different ways to go about developing the patterns that you'll use for making foam armor. First, you can hope that someone, somewhere on the Internet has already made the exact thing you're trying to make and posted all of their patterns somewhere to be easily downloaded. Since that's asking quite a lot, it's not something that should be counted on. The next option is to use aluminum foil, duct tape, and some artful handiwork to develop patterns by hand. Finally, and just as good, a digital model and Pepakura can be used to generate paper templates that will then be traced onto the foam.

Duct Tape Dummy

Sometimes it's hard to find someone who's willing to stand perfectly still while a suit of armor is built around them. A mannequin may seem like a good option to use as a stand-in, but most of them are posed in awkward, unnatural postures and have somewhat nonhuman proportions. But never fear! Just a few rolls of duct tape and an afternoon is all it takes to build a custom mannequin.

Start by gathering up the following tools and materials:

- Six or seven rolls of duct tape

- One roll of plastic wrap

- Trauma shears or bandage shears

- Newspaper or other suitable stuffing materials

- A model who has no issues with claustrophobia

Once you've gathered up your materials, dress the model in something form-fitting (Figure 3-15).

Decide on a pose for the finished piece. This will have to be a position that the model can hold for quite a while. Make sure that the pose will facilitate building the armor without having to cut off the arms and legs later. Once a pose is decided on, wrap the model in plastic wrap, making sure to cover any bare skin and clothing (Figure 3-16).

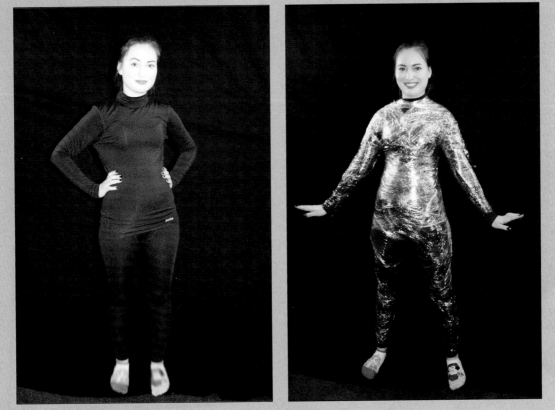

FIGURE 3-15: Have the model wear something form-fitting to keep the proportions correct.

FIGURE 3-16: Plastic wrap

Cover the model completely in duct tape. It's important at this stage to avoid adding a bunch of bulk anywhere. Instead of just trying to mummify the model by unrolling the tape directly onto her body, tear of strips of duct tape and lay them onto the plastic wrap. Keep them as smooth and wrinkle-free as possible (Figure 3-17).

Note that it's important to work quickly at this stage. It may be a good idea to have more than one person layer on the tape in order to minimize the time that the model has to spend cocooned in duct tape.

After the model is completely layered in tape (Figure 3-18), add at least three more layers of tape over her entire body.

The idea is to build up enough thickness for the duct-tape shell to be able to hold its shape even after the model is removed from the inside. If there's any doubt at all, add more tape. When the model is completely unable to move, it's just about done.

After the duct tape shell is re-stuffed, it may be difficult to tell exactly where the model's knees and elbows were. It may be a good idea to mark these key features with a permanent marker or a contrasting color of duct tape as shown in Figure 3-19.

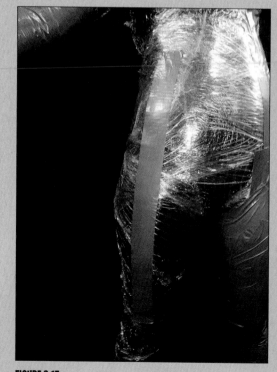

FIGURE 3-17: Laying down smooth strips of duct tape

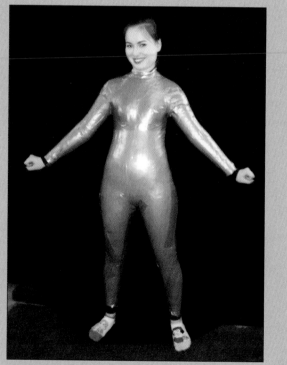

FIGURE 3-18: The model completely wrapped in the first layer of tape

After building up adequate thickness, use the bandage shears to cut open the duct tape, starting at the back of the neck (Figure 3-20).

In order to extricate the model from the duct tape, make relief cuts up the back of the legs and arms (Figure 3-21).

Bear in mind that more cuts mean more seams to tape back together later. Try to keep the relief cuts to the absolute minimum amount necessary for the model to get out.

Once the model has managed to shimmy out of the duct tape, realign the edges of the cut seams and tape them back together. After closing up all of the seams, stuff the dummy with newspaper or expanding insulation foam (Figure 3-22).

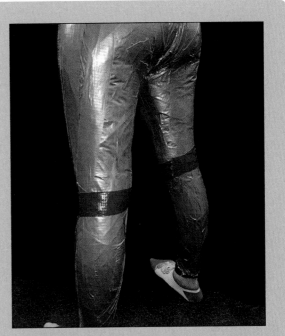

FIGURE 3-19: The red duct tape will make it easier to identify the knee joints on the finished dummy.

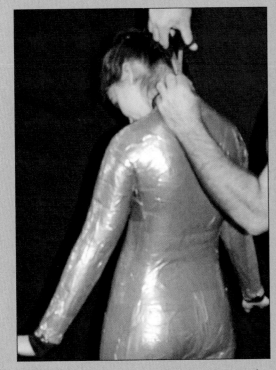

FIGURE 3-20: The moment the model has been desperately waiting for

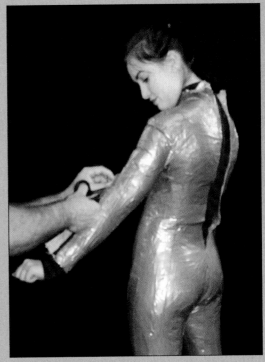

FIGURE 3-21: Cutting open the back of the legs and arms

When you're stuffing the dummy with expanding foam, close off the wrist and ankle holes, then reinforce the seams with extra tape in order to prevent them from splitting.

Fully finished, the stuffed duct-tape dummy is a pretty good stand-in for your model (Figure 3-23).

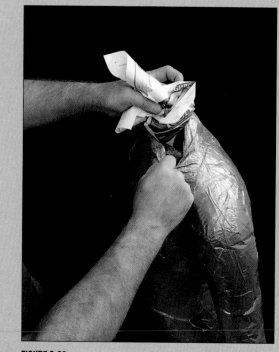

FIGURE 3-22: Pack the newspaper into the dummy as tightly as possible in order to keep everything straight.

FIGURE 3-23: The duct tape dummy can sit still through all of the armor-making while your model goes on her merry way.

Making Patterns by Hand

Let's start with something simple: the shin armor for the unnamed wolf warrior (Figure 3-24).

To get the shape, we just need to decide on a curve for the outer edges. Since the piece in question needs to be symmetrical, start with a piece of heavy cardstock that has been folded in half, draw the desired shape of the outside edge, then cut it along that line and unfold it, as shown in Figure 3-25.

FIGURE 3-24: The armored greaves of . . . is the name Ethlwulf okay? No?

FIGURE 3-25: Making the template for the shin armor

This template can now be used to do a quick test-fit against the wearer before tracing it onto the foam. Once the shape is traced onto the foam, simply cut out the two pieces (Figure 3-26).

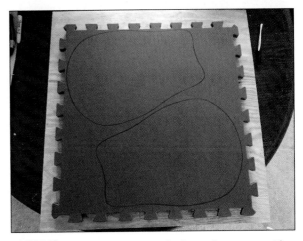

FIGURE 3-26: The parts traced onto the foam, then cut out with a knife before sanding the edges smooth

> ## 🔫 Maker Note
>
> In this case, both shins are the same shape and each piece is symmetrical. This means that the two pieces are essentially interchangeable. If the inner and outer edges were not the same shape, the two pieces would probably still be mirror images of each other. Once a template is made for the left side, it can then be flipped over and used as a template for the right side.

Once the foam pieces are cut out, all that's left to do is cut a vertical groove into the backside along the centerline, then glue the groove shut to form a tightly rounded corner on the outside. Once that's set up, use a bit of heat to curve the sides, as needed, to form a good fit (Figure 3-27).

FIGURE 3-27: Masking tape holds the pieces in position while you're waiting for the glue to cure and the foam to cool.

That's all that's needed in order to get the general shape of the greaves. The next step will be to add trim and other details, but we'll get to that later in this chapter.

For now, let's step up our game a bit and make something a bit more complicated: the gauntlet for our still-unnamed wolf warrior (Figure 3-28).

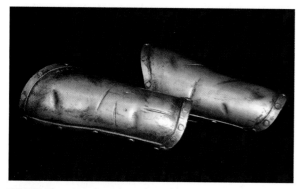

FIGURE 3-28: We could call her Mysandria . . .

To get your basic shapes worked out, start by gathering the following tools and materials:

- Aluminum foil or plastic wrap
- Duct tape
- Bandage shears or safety scissors
- Permanent marker
- The foam that the armor will be made of

Start by wrapping the body part to be fitted (Figure 3-29).

FIGURE 3-29: The forearm wrapped in foil to keep the tape from sticking

🔫 Maker Note

Unless the wearer has some significant symmetry issues, both arms will probably be pretty close to the same size. Still, if the parts are supposed to match from left to right, it's a good idea to make the template off of the wearer's dominant side (i.e., the right arm for a right-handed person). If the template is made to fit snugly on the weaker side, it may be too tight to fit on the stronger side.

Cover the foil completely with one layer of duct tape (Figure 3-30).

Use a permanent marker to draw lines where the tape will be cut (Figure 3-31). Make a tick mark across the cut line every inch or two (2–4 cm).

FIGURE 3-30: Tape covering the foil to keep it from deforming

🔫 Maker Note

If the armor design calls for seams or trim, these make ideal places to split the pattern up. In this case, the gauntlet has a piece of trim that runs along either side of the forearm, so we'll be cutting the pattern there.

These will help to realign the parts on either side of the seam later.

Using the bandage shears in order to avoid damaging the body part underneath, cut the foil/tape mess along the cut line. Then cut notches in the edges where the tick marks were, as shown in Figure 3-32.

This is almost the exact shape to cut out of the foam, but not quite yet. This shape fits perfectly around the forearm it was made on and that's great. But it's also just a thin layer of foil/tape weirdness. Unless the foam that the armor will be made of is just as thin, there will have to be some additional width added to the template in order to account for the girth added by the thickness of the foam. Also, since this one-piece gauntlet will need to slip on over the wearer's hand, it'll have to be made even bigger in order for that hand to fit through.

To determine the additional width, make a long strip out of foam and wrap it around the hand of the wearer it was sized for, as shown in Figure 3-33.

After marking the length of the foam strip, compare it to the width of the wrist area on the tape and foil template (Figure 3-34).

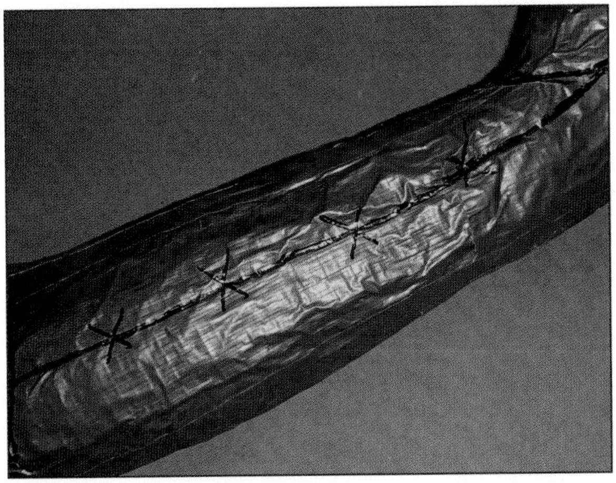

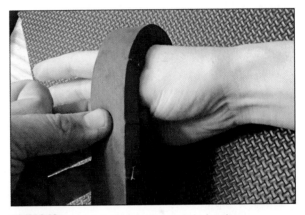

FIGURE 3-33: Foam strip wrapped around the hand

FIGURE 3-31: The cut line with tick marks

FIGURE 3-32: The template with alignment notches cut along the edges

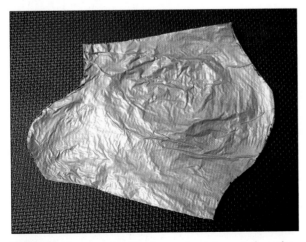

FIGURE 3-34: Huge difference

At this point, modify the foil template by widening it until the wrist area is the same width as the length of the foam strip that wrapped around the wrist. In this case, that means we'll need to add almost 2" (4.5 cm) by simply splitting the template and adding a bit more tape in the middle (Figure 3-35).

Now the template can be traced onto the foam, as shown in Figure 3-36.

Cut the parts out of the foam and fit them together with masking tape in order to ensure that there are no problems where the edges are supposed to connect to each other. This is also a good time to test-fit the parts to the wearer and make sure that nothing has gone wrong in the process of developing the template (Figure 3-37).

At this point, these taped together gauntlets look nothing like the reference image that we started with. Now it's time to add some finesse by heating them up and bending them into shape. Specifically, we'll start by adding a bit of convex shape to the area that will correspond to the back of the hand (Figure 3-38).

FIGURE 3-35: Widening the template

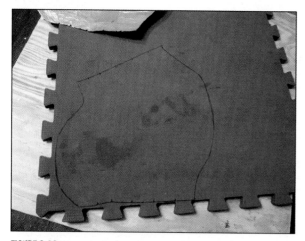

FIGURE 3-36: The template traced onto the foam

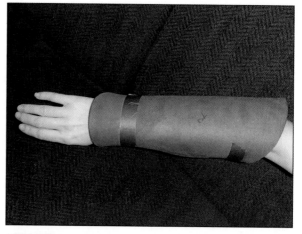

FIGURE 3-37: The test fitting

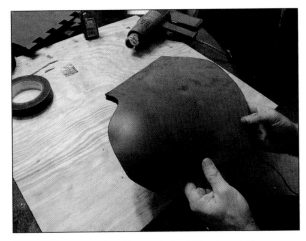

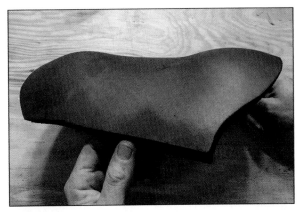

FIGURE 3-39: More heating, bending, and stretching to get the rounded shape for the elbow area

FIGURE 3-38: Heating, bending, and stretching to get the rounded shape for the back of the hand area

Then the same thing is done for the meaty area inside the forearm around the elbow (Figure 3-39).

After getting the bulbous areas shaped, it's time to heat up the rest of the parts and put a bend in them so the seam can be glued. Use some strategically placed strips of masking tape to hold the seam closed until the glue has a chance to dry (Figure 3-40).

Now that we have the basic shape put together, all it needs is a few details and some

battle damage—more about that later. First, let's look at making an even more complicated shape: the breastplate (Figure 3-41).

For reasons that should be somewhat obvious, form-fitting female chest armor presents a unique set of challenges. There are compound curves and corners to deal with everywhere. Still, we'll start by creating a foil-and-tape pattern in much the same

FIGURE 3-40: Masking tape to hold the seam while the glue dries

FIGURE 3-41: Armored boobs are a staple of science fiction and fantasy character design.

way as we did for the gauntlets. In this case, we'll start by wrapping foil over the chest of a duct-tape dummy, as shown in Figure 3-42.

After wrapping the chest tightly in foil, apply a single layer of duct tape (Figure 3-43). Again, the idea here is to reinforce the easily torn foil shape without adding any unnecessary thickness. If you use too much tape, you'll cause distortion along the way.

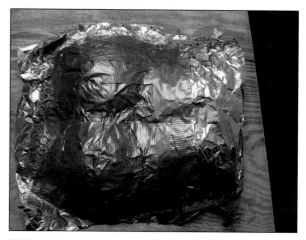

FIGURE 3-43: A single layer of duct tape laid over the foil with minimum overlap

Now's the time to remove the foil-tape shape from the chest and cut it so that it lays flat. In this case, the piece will end up being built much like a bra, with a simple curve that conforms to the ribcage under the bust and a pair of cups that conform to the shape of the breasts. We can start by cutting off the bottom portion along the under-bust line, then separating the cups by splitting them right down the center of the sternum. This makes the pattern into three pieces, as shown in Figure 3-44. Start by drawing cut lines with a permanent marker. Include tick marks every inch or so (about every 2 cm). Then cut the pieces apart.

FIGURE 3-42: Duct-tape chest wrapped in foil

Maker Note

It may be tempting (and entertaining) to try to use the chest of the actual model for this step. Unfortunately, unless your model has rock hard skin and can hold their breath for the duration of the entire process, using a live model will introduce a lot of errors and deformities along the way. You're really better off using a duct-tape dummy or a life-cast copy of the model's chest.

FIGURE 3-44: Separating the three separate pieces of the breast-plate will be based off of just two patterns because symmetry is our friend.

Unless the model is some sort of mutant or plastic surgery addict, female breasts are almost always asymmetrical. In most cases, one will be ever so slightly larger than the other. This is normal. Typically, the larger one will correspond to the non-dominant hand (i.e., the left breast will be ever so slightly larger on a right-handed woman). For the final piece to be symmetrical, discard the template for the smaller, dominant side, and instead use the template for the larger, non-dominant side.

The piece that wraps along the ribcage should be fairly easy to flatten out without stretching or tearing it. Now is the time to determine the exact shape of the bottom edge of the breastplate, as shown in Figure 3-45. After trimming the foil/tape piece to the ideal shape, fold it in half to verify that it's symmetrical. Then transfer the outline to a piece of paper that's been folded in half. After cutting it out and unfolding it, the paper template will be just the right shape.

Now it's time to move on to the cup shape. Since a round cup shape will not want to lay flat, it will need to be cut in at least one place

(Figure 3-46). These cuts are analogous to the darts that are used by seamstresses when making form-fitting garments without folds or gathers. For deeper or curvier shapes, more cuts may be needed.

> ### 🔫 Maker Note
>
> Everywhere that there was a tick mark on the cut line, you need to cut a notch into the edge of the paper template. These notches will become alignment marks later when it's time to put the foam pieces together.

FIGURE 3-46: A cut from the middle to the edge of the cup will allow it to lay flat.

> ### 🔫 Maker Note
>
> Since the cuts can be placed anywhere, try to choose areas that will later be covered by trim or other details. That will make it easier to hide the resulting seams.

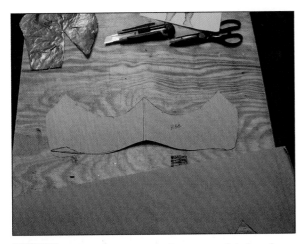

FIGURE 3-45: Making the symmetrical paper template based on the shape of the foil/tape pattern

After making enough cuts to get the foil/tape piece to lay flat, trace the design onto a piece of paper (Figure 3-47), making sure to transfer the locations of all of the tick marks. Then cut the template out, making notches along the edges where the tick marks were.

Trace the paper templates onto the foam and cut out the pieces (Figure 3-48).

Before gluing anything together, start by heating the foam pieces with a heat gun and shaping them by hand. While the shape can be fine-tuned once the pieces are glued together, it's best to get it as close as possible while they're still separate in order to avoid messing up the glued seams. Start by bending the rib piece to roughly match the shape of the wearer's ribcage, as shown in Figure 3-49.

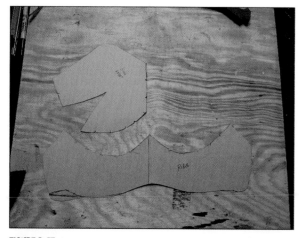

FIGURE 3-47: Creating a smooth paper template from the foil/tape version

FIGURE 3-48: Cutting out the foam parts based on the paper templates

Maker Note

Since the cups will be mirror images of each other, the template for the right side can be flipped over and used as a template for the left side.

FIGURE 3-49: The curved rib piece

Next, heat the cup piece and shape it over a knee or ball in order to get the proper curve to fit the wearer's breast (Figure 3-50).

After the parts have been shaped, it's just a matter of gluing the seams together (Figure 3-51).

For more complex shapes, it's simply a matter of starting with templates like these and building mockups of the necessary pieces out of card stock.

Then disassemble the cardstock mockup and use the pieces as templates for the build.

FIGURE 3-50: The shaped breast cup before gluing

FIGURE 3-51: Assembled base shape for the breastplate

Final Detailing

Now that the basic shape has been made, it's time to bring it to life by adding all of the little details that made the character cool to begin with. These are the seams, rivets, dents, scratches, and raised panels that really make things real.

Seams and Panel Lines

To make a seam line that runs across a nice, smooth, single piece of foam, start by marking the line where the seam is going to be. Then use a sharp knife to barely cut into the surface of the foam. At this point, it's probably not visible. Now use the heat gun to slightly cook the surface of the foam. As the foam is heated up, the surface will contract slightly, opening up the cut groove and making a very visible seam line, as shown in Figure 3-52.

Dents and Scratches

There are two basic ways to add dents. The first is to grind into the foam with a file or coarse sandpaper. Then go back over the area with a heat gun or blowtorch in order to seal the surface and prep it for paint. This creates a thoroughly believable gouge in the surface (Figure 3-53).

The other option is to heat the tip of a metal tool and press it into the foam. Just be careful not to use too much heat or the foam will melt into a gooey mess.

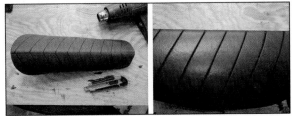

FIGURE 3-52: Making a seam line that's only skin deep

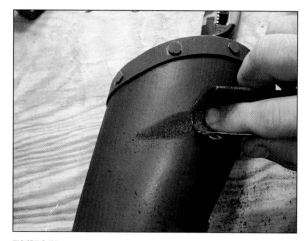

FIGURE 3-53: Ground-in dents

For scratches, the best bet is to carve them in with a file, sand around the edges so they don't look too manufactured, then heat the area to seal the surface before painting it. The end result is shown in Figure 3-54.

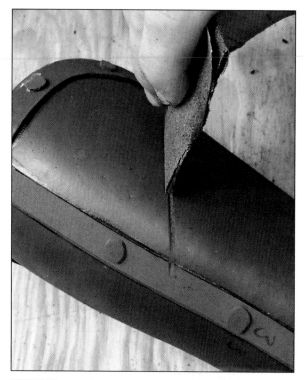

FIGURE 3-54: Scratches carved with a file and then heat-sealed

Raised and Layered Details

When it comes time to add small raised areas or layered details, there's no better option than thin craft foam. It can be cut into strips to create trim or borders like those shown on the gauntlets in Figure 3-55, or the raised diamond pattern on the breastplate in Figure 3-56.

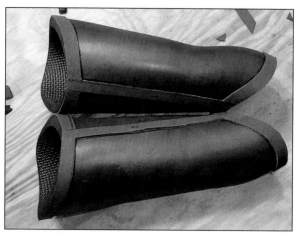

FIGURE 3-55: Gauntlet trimmed with a strip of craft foam

FIGURE 3-56: Raised breastplate pattern made of diamond shapes cut from sheets of craft foam and trimmed with strips of more foam

Rivets and Screw Heads

The best thing you can use to simulate the visible heads of fasteners, such as rivets and screws, is real rivets and screw heads. They're readily available at any hardware store and, as an added bonus, they can be used to attach any necessary straps to the armor pieces.

If you don't want to use real hardware, no problem. You can fake a rivet detail easily by using your rotary tool fitted with a sanding drum. Simply turn the speed up high, and touch the spinning end of the drum to the foam surface. The friction will melt a neat little circle into the surface of the foam that'll look just like a rivet head (Figure 3-57).

Prepping for Paint

Look closely at the nice, smooth surface of the foam armor. Notice it's not smooth at all. In fact, the entire surface is covered with countless tiny holes. What madness is this? It should come as no

surprise that the foam is essentially just a bunch of rubber bubbles stuck together. This means that unless the surface is heat-sealed to melt these bubbles together, there will be at least a few holes visible on the surface.

But never fear. In their infinite wisdom, someone invented a spray-on rubber coating that makes an ideal primer for this surface: Plasti Dip rubber coating. Available in most hardware stores, it can be purchased in a wide variety of colors and dries to form a layer of soft, flexible rubber.

Sure, they may have intended for you to use it to add non-slip coatings to tool handles and whatnot, but deep down inside, they had costume and prop builders in mind.

To seal the outside of the foam, simply spray a couple of light coats onto the porous surface. It will absorb the paint, and the finished product will have a dull, gray-ish finish. If there are still areas that seem porous, allow the Plasti Dip to dry, and spray

FIGURE 3-57: Simulating a rivet head with the tip of a sanding drum on a rotary tool

on another coat. Repeat, as necessary, until you've achieved the desired surface smoothness.

Another option for sealing the surface is to use Mod Podge. This is a commonly available product intended for use in decoupage. It also works in certain applications as a glue, finish, and (just what we're after) a sealer. It can be brushed or rolled onto the surface and will provide a durable shell once dried. While it can often be tougher to get it to lay flat like Plasti Dip, brushing it on with a bristle brush can make for a great wood grain texture, as shown in Figure 3-58.

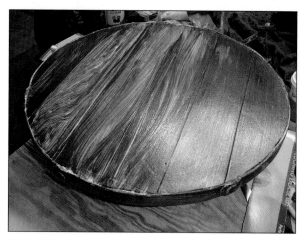

FIGURE 3-58: The wolf warrior's shield coated in Mod Podge

Paint!

Once the surface is sealed, all that's left to do is coat everything with color to hide the fact that it's all made of foam rubber, as shown in Figure 3-59. The main consideration here is to use paint that's a bit flexible. That way, it'll give a little bit as the foam bends and moves instead of cracking and creasing as you walk around. Latex and acrylic paints are ideal for painting foam.

How was all of this done? More on that in Part V of this book.

FIGURE 3-59: Fully painted, this lady's armor really comes to life!

3D PRINTING AND CNC CARVING

Let the Robots Do the Work

OVER THE COURSE OF the last decade or so there has been a lot of hype surrounding 3D printers and other emerging "rapid prototyping" technologies. At the same time, the development of ever-cheaper computer numeric control (CNC) milling and carving machines for in-home use means there are even more options for automated manufacturing in the home. The problem is, for most normal people, there really isn't all that much use for keeping any kind of 3D printer or CNC machine in the home.

But that's normal people. For prop makers, it's a completely different story. The development of these new technologies almost means that the last four chapters can be completely ignored. After all, there would seem to be no reason to spend a couple of nights putting together a Pepakura model when you can have your desktop 3D printer spit out the pieces while you sleep. You'd be silly to spend all day stacking up layers of plastic to build a prop ray gun or an axe if you could just as easily program the design into your three-axis CNC machine, and watch a movie while the machine carves out the parts.

The important thing to keep in mind, though, is that these automated rapid prototyping systems are not "magic button" solutions. While they will make a few processes easier or faster or cleaner, they won't do *all* of the work for you. You'll still need to come up with the original digital models that the machines will either carve or grow parts from. Even the very best machines available to consumers today will make parts that require at least some cleanup or processing work. And even though they're often referred to as *rapid prototyping* systems, they still take time to make things.

While there are a lot of different machines that are helpful in the modern workshop, let's take a look at the two you hear the most about lately: 3D printers and CNC machines.

The main difference between the two is this: 3D printers use some sort of additive process to put together layers of material until an object is made, whereas CNC machines use a subtractive process to remove excess material from a piece of stock until only the desired object remains. The details of the process will vary from one machine to the next, but that's the difference in a nutshell.

So which one is better? It really depends on the project in question. They both have strengths and weaknesses. When it comes to commercially available machines for hobbyists (and speaking in wild generalizations), CNC machines tend to be able to make things faster and cheaper, while 3D printers tend to be able to make things with smaller details.

So if you're working on something large with a lot of small details, it's wise to take advantage of both types of rapid prototyping. As an example, take a look at this digital render of our wolf warrior woman in Figure 4-1.

All of the smooth curves, fine details, and organic shapes would make this a huge nightmare for a Pepakura or foam build. But, if you've got a high-resolution 3D model, it's a simple matter of making sure that the mesh is watertight and delegating the manual labor to the machines. In this case, the model is designed with the ears as separate pieces from the rest of the helmet. 3D printing the main portions of the helmet would take forever. The ears alone, however, are only about 20 hours' worth of work for a hobby-level fused deposition modeling (FDM) 3D printer.

Since the main body is much bigger, growing it with a 3D printer would take a ridiculously long time, and, for most machines, it would have to be made in a lot of separate parts. A CNC machine can really speed up the process of making the prototype. Using a variety of materials ranging from metal, wood, or even rigid foam, parts can be

FIGURE 4-1: We could call her *Ylestra, Warrior Princess of the Wild North* (original ZBrush model by Peter Rubin).

carved out relatively quickly. For example, it took about 10 hours for a three-axis CNC machine to make the parts shown in Figure 4-2.

The surface finish of the CNC-carved parts will vary depending on a wide variety of factors, including the size and shape of the cutting bit, the type of material being carved, and the speed of the bit as it travels along the cutting path. In this case, the surface is a bit fuzzy where the medium-density fiberboard (MDF) is cut. This fuzziness is apparent in Figure 4-3.

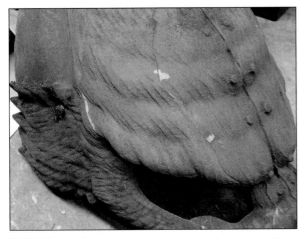

FIGURE 4-2: Wolf helmet parts whittled out of readily available, inexpensive MDF

FIGURE 4-3: The MDF surface with a coat of red primer is still fuzzy.

After a few more coats of primer and a tiny bit of sanding, it only takes a bit of spot putty to get rid of the last of the fuzzies (Figure 4-4).

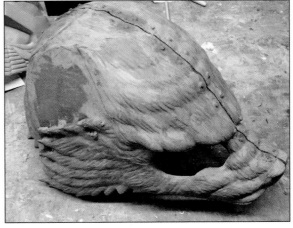

FIGURE 4-4: Smoothing the fuzzies with spot putty

As usual, another coat of primer does a good job of letting you know if there are areas that need more work (Figure 4-5).

After a couple of rounds of sanding, filling, and primer, the helmet will be ready for a final coat of gloss paint (Figure 4-6).

The separate ears also need a bit of cleanup before they're ready for paint.

FIGURE 4-5: Gray primer to show that the whole thing is nearly smoothed out

FIGURE 4-6: Smooooooth!

With most 3D printers, there will be discernible striations, or *build lines* where the different layers were stacked up in the vertical axis. As the technology improves, these lines are getting less and less noticeable. For higher-quality printers, it's possible that these lines can be virtually undetectable after a couple of thick coats of primer. For lower-resolution machines, they may need to be filled in with spot putty and then sanded until smooth. If that sounds like too much work, there are a few manufacturers that offer self-leveling epoxy resin designed specifically for this purpose (Figure 4-7).

After allowing them to cure, the ears are ready for a nice, glossy coat of paint, as well (Figure 4-8).

Why is it painted that color? If the plan is to mold and cast these parts, the actual color doesn't matter at this stage, only the shiny, smooth surface. The problem with making the helmet out of MDF is that it will be heavy and a little bit fragile. So, unless it's intended to be used for display only, a better idea would be to pull a mold off it; cast a sturdier, lightweight copy; and wear *that* one around. We'll explain how to do all of that in Part II, "Molding and Casting."

FIGURE 4-7: Using a specialized self-leveling resin to fill the build lines for the 3D printed ears

FIGURE 4-8: It's actually a new color; a sort of light red.

PART II

Molding and Casting

MOLD-MAKING BASICS
Leave a Lasting Impression

THERE ARE MANY REASONS to make a mold and cast a copy of something, but they usually boil down to one of the following two: The first is that the original may be made of something inappropriate and needs to be reproduced with a different material. The second is that you may want to make lots of duplicates without repeating the tedious or expensive process involved in creating the original.

Consider these examples.

Suppose you're really good at sculpting things out of clay, and you have just spent a week sculpting a perfect replica of a character's signature pair of gauntlets. You can't really wear a pair of clay gauntlets, at least not for very long. Once a mold has been made, however, those same gauntlets can be reproduced in a variety of sturdy materials that can be sanded, painted, drilled, and so on. What's more, if the reproductions break, remaking them is just a matter of pulling another copy out of the molds.

Now suppose you have a big group of friends, and you want to dress them all in matching suits of armor. Instead of spending weeks rebuilding the same Pepakura models over and over until you go blind, bleed to death from paper cuts, or overdose on cyanoacrylate fumes*, you can make molds of the parts of the first suit, and build all of the matching suits you could ever want. If the mold wears out, you can simply make a new mold of one of the cast parts.

---

*This may or may not be possible. We're not doctors!

Mold-making requires some degree of skill and no small expense, but with some supplies that can be found at better art supply or hobby stores (or through countless websites), it only takes a bit of time and practice to master the basics.

The Simplest Molds Ever

Believe it or not, we've almost all made some kind of mold at some point in our lives. Chances are, it was some sort of kindergarten arts-and-crafts project when you were too small to be trusted with a pair of scissors. Figure 5-1 shows a prime example.

That's a bit of clay with an impression of a five-year-old's hand pushed into it before glazing and firing. Now, imagine if something like plaster was poured into that impression and allowed to cure. As long as the plaster didn't bond to the clay, it could be (very carefully) removed and would end up being a pretty faithful re-creation of that little hand.

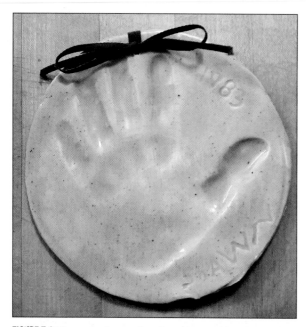

FIGURE 5-1: The author's very first foray into mold-making was way back in the 1900s.

Another example of a mold most people might not realize they've already made probably hundreds or even thousands of times is a footprint. If someone came along behind a person strolling barefoot on the beach and filled the footprint with plaster, they'd have a pretty reasonable copy of the foot. Another example from the beach: using a small plastic bucket as a mold to cast a tower for a sand castle.

As you can see, most of us have been making molds and castings in some form or another since before we could reliably use the bathroom without supervision. Now, it's just a matter of you refining those skills to make things that are little more complicated, and doing it a bit more reliably.

Mold-Making Material

Before you begin making molds, you must determine what the finished castings need to be able to do. Will the finished part be set on a shelf for display? It'll need to be made out of something that will look good once it's painted up. Will it be something that is going to have to be knocked around and abused? Perhaps it'll be better to cast it out of a rubbery material that's more likely to bounce than shatter.

In any case, conventional wisdom teaches that rigid castings usually require a soft, flexible mold, and flexible castings can be made from a rigid mold.

This used to mean that prop makers would have to spend a lot of time learning to use a wide variety of materials to make molds, depending on what they were making. Since most of these materials were repurposed construction materials, which were designed with adhesive qualities, it was necessary to match molds to casting materials that they wouldn't stick to. Even then, it was still a good idea to use a release agent to keep the molding materials from sticking to the prototypes, and later to keep the casting materials from sticking to the molds. Using too much release agent could cause some of the details to get lost in the casting process. Using too little release agent, or leaving it out altogether, would precipitate a day-ruining catastrophe with parts sticking to molds and both of them being destroyed in the process.

Enter silicone rubber.

With the increasing availability of silicone mold-making rubbers, life has become much, much simpler for mold-makers. Silicone can cure at room temperature (often referred to as *room temperature vulcanizing*, or *RTV*) and, aside from perhaps a scale to get the mixing ratios correct, using it requires no specialized equipment.

Silicone can be purchased in countless formulations with faster or slower cure times, higher or lower viscosity, and harder or softer cured properties. Various additives can be mixed in to alter all of these properties as well, making it even more versatile. There are silicones that can be used against the skin to duplicate body parts, and there are silicones that are food-safe to allow you to cast parts in ice, chocolate, or whatever you like. It even comes in different colors.

When it comes to detail reproduction, silicone rubbers have no equal. The tiniest thing on a prototype will be faithfully reproduced on all of the castings. As a matter of fact, paleontologists use silicone molds to accurately reproduce fossil samples right down to their microscopic cellular structures!

The best part: with a few rare exceptions, the only thing that will bond to silicone rubber is more silicone. This means that a release agent, while it might be helpful, is not actually necessary.

There are still occasional applications that make other molding and casting materials a good idea, but for our purposes, there's usually no better all-around mold-making material.

All of these wonderful advantages do come at a price. At the time of this writing, some of the most common mold-making silicones will cost anywhere from $80–$200 per gallon, but with a little bit of experience, you can take full advantage of all that silicone has to offer and minimize waste along the way.

Mold-Making Terminology

Before we dive into mold-making in the next chapters, it's a good idea to know the lingo. As with any group of skilled artisans, mold-makers have invented themselves a whole separate language over the years to describe the tools, materials, and processes involved. But, despite all of their secret code words, it's not some kind of rocket surgery. Here are a few key terms you'll hear mentioned in the next handful of chapters.

CASTING   Also known as a *part*, this is the thing that is made in the mold, a duplicate of the original model.

CATALYST   Also called a *curing agent*, this is a substance that causes (or speeds up) the curing of a compound when added.

CURING TIME   The period of time required for a material to fully solidify. Also known as *demold time*.

DEMOLDING The process of removing a mold from a model or a casting from a mold.

DOE A deer. A female deer. Not at all related to mold-making.

DRAFT The slope or taper given to the vertical surfaces of shapes, when necessary, to make it easier to remove them from a mold.

FILLER An inert material that is added to a mixture to reduce cost, reduce weight, modify mechanical properties, or to improve the surface texture.

GANG MOLD A mold with two or more mold cavities or impressions (see Figure 5-2). Essentially, a single mold that produces more than one casting per pour.

GEL The semisolid phase that develops during the formation of a resin from a liquid.

GEL TIME The amount of time after introducing a catalyst before the material turns from liquid to an unworkable mass. Also known as *working time*.

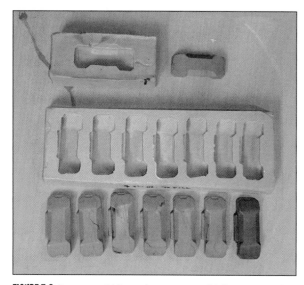

FIGURE 5-2: A gang mold is used to create multiple castings of the same object simultaneously.

KEYS Unique shapes that force mold halves or sections into alignment when closing or joining together so they will not shift. Typically, this is some sort of protrusion in one part with a matching depression in another part.

MODEL Also referred to as the *prototype* or *master*, this is the original object to be reproduced.

MOLD The cavity or form that has a negative or reverse impression of an original model. Molds can be made of a rigid material, such as plaster or plastic resin, or, more commonly, a flexible material such as rubber.

MOTHER MOLD A rigid shell made to support a flexible rubber mold.

PARTING LINE A mark on a molded piece where the sections of a multi-piece mold mate together when the mold is closed.

RELEASE AGENT A material applied in a thin film to the surface of either an original model prior to making the mold, or the mold surface prior to casting. Release agents prevent adhesion between two materials that would otherwise stick together.

ROTOCASTING The process of rotating a mold to coat the inside surface with a thin layer of casting material. Also known as *slush casting*, *spin casting*, or in more formal settings, *rotational casting*.

SPRUE A funnel-shaped opening in a mold where you pour a casting material into the mold. Also, the waste casting material left over when opening a mold after you make a cast product.

UNDERCUT   Any indentation or protrusion in a shape that will prevent its withdrawal from a one-piece mold.

VENT   A small tube built into a mold to allow air bubbles to escape.

VISCOSITY   The resistance of the material to flow. A material with low viscosity will flow easily. A material with higher viscosity will not flow as easily. As an example, ketchup has a higher viscosity than apple juice. That's why it's harder to get ketchup out of the bottle.

FLUX CAPACITOR   Not actually a thing. Just checking to see if you were still awake.

ONE-PIECE MOLDS
You've Got to Start Somewhere

THE SIMPLEST TYPE OF mold is a one-piece mold, also occasionally referred to as a *block mold* or *box mold*. This type of mold is used to replicate small parts that can easily be contained during the molding process.

Easy Version: A Part with a Flat Backside

This first example of a box mold is ideal for use when you need to replicate something with a flat side that has no details. When laid on the side, the piece has to have no significant undercuts where bubbles will be trapped.

An ideal candidate for this kind of reproduction is Hunter's belt buckle. Let's begin by taking a look at the prototype in Figure 6-1.

To start, it needs to be sealed so that the silicone won't drool its way into the nooks and crannies. The silicone won't glue itself to the surface, but it can seep into any exposed pores or seams. These are parts where the rubber may tear during

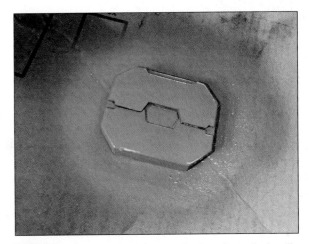

FIGURE 6-1: Hunter's belt buckle: a simple piece that nobody will see the back of

demolding. So, after a couple of coats of primer to seal up the surface, the whole thing gets a nice, shiny coat of gloss paint, as shown in Figure 6-2.

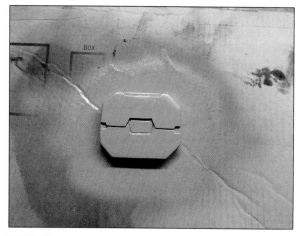

FIGURE 6-2: Shiny! At this stage it doesn't matter what color the prototype is.

Once the paint has had plenty of time to dry, it's time to set it up for molding.

Step 1

Glue it down to a piece of cardboard so it won't float away once the silicone is poured over it (see Figure 6-3).

FIGURE 6-3: Prototype securely anchored to a piece of cardboard

Step 2

Glue strips of cardboard to form watertight walls around the prototype as shown in Figure 6-4. The box should allow for at least a ½" (approximately 1 cm) gap all the way around the outside of the prototype. It should also be at least 1" taller than the highest point on the prototype.

FIGURE 6-4: A box

You can use any kind of glue you're comfortable with as long as it will be strong enough to hold the box together. Hot glue is an ideal option because it's cheap, fast, and easily available. Unless you develop some sort of odd emotional attachment along the way, the mold box will be discarded in the end, so there's no sense in making it from something expensive.

Step 3 (optional)

Before pouring in the rubber, spray a thin layer of release agent onto the prototype. Silicone doesn't typically bond to anything other than silicone, but you don't want to find out the hard way what the exceptions are.

Step 4

After double-checking to be sure that the box won't leak, mix up a batch of silicone RTV mold-making rubber in accordance with the manufacturer's instructions (Figure 6-5). The silicone usually comes in two parts that are different colors. If there are still streaks of different color, you're not done mixing. Be sure to scrape the sides and bottom of your mixing cup in order to ensure that there's no unmixed residue clinging to them just waiting to cause problems.

Step 5 (optional)

If you have access to a vacuum chamber, now's the time to use it to de-gas the silicone. This means placing the mixed liquid rubber in a vacuum, which will cause the bubbles trapped in the rubber during the mixing process to expand and burst, leaving the silicone mixture free of air bubbles that can cause minor flaws in the cast parts. Hopefully your machine is one that lets you watch all of the bubbling and bursting. It's fun! Just make sure that the silicone is in a container that will allow it to expand as much as three times its original volume before settling back down. If you don't have a vacuum chamber, you can get by without it.

Step 6

Pour the liquid rubber into the mold box (Figure 6-6).

FIGURE 6-5: Properly mixed silicone is all one uniform color.

FIGURE 6-6: The Pouring

When pouring the rubber, the most important thing is to avoid trapping any air bubbles against the surface of the part. Here are a few tips:

- Start by pouring into a low part of the mold. This way the rubber will flow up and over the part in the mold without trapping as many bubbles against the surface of the part.

- Be sure to fill the mold box so that there is at least ½" of silicone over the top of the part. That way, any bubbles will have a chance to rise up and away from the surface of the prototype. A bit more is even better.

- Be patient. Every mistake in this stage will cost extra time in the casting and painting stages. It's better to fix one problem now instead of dozens later.

Step 7

Leave it the hell alone. Find someplace where the mold can sit undisturbed at room temperature for at least as long as the manufacturer's stated cure time for the material. It's also not a bad idea to cover it with something, since every suicidal insect within flying, crawling, or falling distance will suddenly decide that your wet, sticky silicone is the best place to end its sad insect life (Figure 6-8).

FIGURE 6-8: Frank the fly has finally found his final resting place. He annoyed everyone he ever met, but he still deserved better.

The Poor Man's Vacuum Chamber

If you don't have access to a vacuum chamber and you're still worried about bubbles being trapped in your silicone, don't fret. Remember: bubbles rise, so all that's needed is to punch a hole in the bottom of the mixing container that's big enough to allow the silicone to leak out and into the mold box (as shown in Figure 6-7). This will allow the bubbles to continue to rise to the top of the mixing container, while the de-gassed, bubble-free rubber dribbles out of the bottom.

By holding the container higher above the mold, the silicone will stretch into a long, thin stream, and what few bubbles might remain will pop along the way.

FIGURE 6-7: An inexpensive de-gassing plan

Step 8

Once the rubber has completely cured, flip the mold over and remove the cardboard (Figure 6-9).

Step 9

Peel the rubber off of the prototype (Figure 6-10). This is called *demolding*.

Step 10

Make sure there aren't any bits of the prototype left behind in the mold.

FIGURE 6-9: A mold without a box is still a mold.

FIGURE 6-10: Peeling the mold off the prototype

🔫 Maker Note

It's a good idea to think of it as peeling the mold off the prototype instead of prying the prototype out of the mold. In many cases, you'll be casting rigid materials in these flexible molds, and, if the parts are thin, they may break if they're pried out. If they're left flat, bending the rubber won't cause any breaking strain on the parts.

Step 11

Determine how much resin you'll need to fill the mold. There are a few easy ways to do this.

OPTION 1 (MESSY)   Use a graduated mixing container that's partly filled with water. Note the total volume of water in the container, then submerge your prototype in it (Figure 6-11). The added volume of the prototype will cause the water to rise. Note the total volume and subtract the original volume. The difference between the two is the volume of the piece to be cast.

OPTION 2 (LESS MESSY)   Fill the mold with dry rice or some other material that won't leave a residue inside the silicone mold cavity (Figure 6-12). Once it's full, pour the rice out of the mold into

FIGURE 6-11: Using the prototype and a measuring cup to determine casting resin volume

a measuring cup. The volume of rice is about the same as the volume needed to fill the mold.

FIGURE 6-12: Using rice to determine casting resin volume

OPTION 3 (PROBABLY NOT AT ALL MESSY) Do the math. Measure the prototype's length, width, and height. Multiply these together to get the volume of the part. Surface details such as bumps or depressions will increase or decrease this volume, so adjust as necessary. You'll probably get close enough.

Step 12

Mix a batch of casting resin in accordance with the manufacturer's instructions.

🔫 Warning

Like most plastic products, urethane casting resins are petroleum-based. When they cure, they will often have a tiny bit of oily residue rise to the surface. This means that the part should be cast in a single pour. If you misjudge the amount of resin needed and have to add more, do not delay. If you wait for any length of time, this oily residue will form at the surface and act as a sort of release agent. It may be impossible for subsequent layers of resin to bond with this residue in place. You've been warned.

Step 13

Allow the resin to fully cure (Figure 6-13). Bear in mind that these types of material will cure via an exothermic chemical process. This means that it will generate heat. The bigger the part, the more heat it will generate. The more heat, the faster the cure. For small parts, this usually doesn't matter all that much. For large parts (or large batches of resin for a lot of small parts), it can become a significant problem. Plan accordingly.

FIGURE 6-13: As the resin cures, it will slowly change color. Fun!

Step 14

Demold the casting (Figure 6-14).

FIGURE 6-14: Just as we peeled the mold off the prototype, so too we must peel the mold off the cast parts.

Step 15

Gaze proudly upon the masterful bit of plastic you have wrought, which you can see in Figure 6-15.

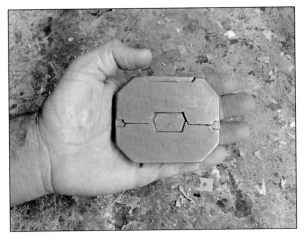

FIGURE 6-15: A cast part made out of plastic, science, brawn, and know-how. Marvelous!

Step 16

Repeat steps 12–15 as needed to get as many copies of the part as your heart desires (Figure 6-16). Depending on the quality of the silicone rubber you use, you can get dozens of castings out of a mold like this.

FIGURE 6-16: Belt buckles everywhere!

Cold Cast Metal

Want your parts to look like metal without having to learn about smelting, or burning your house down? While there are a lot of techniques that can be used to simulate the look of real metal parts, one of the best methods is to actually use real metal. You can "cold cast" metal parts using fine-ground metal powders that can be purchased at art supply stores or most of the same places that sell casting resins.

Step 1

Dust the inside of mold with whatever metal powder you've decided to use (Figure 6-17). Use a small brush to make sure that it's evenly distributed over the surface of the mold, and then pour out any excess. You don't want it piled up anywhere.

FIGURE 6-17: Dusting the mold with powdered brass

Step 2

Mix a batch of resin as you normally would.

Step 3

Stir a bit of the metal powder into the mixed resin (Figure 6-18). The exact amount is a matter of personal preference. More will produce a more convincing metallic surface.

Step 4

Pour the resin/metal powder mix into the mold.

Step 5

Wait for the resin to cure, then demold as usual.

Step 6

Buff the surface of the part gently with some fine steel wool (Figure 6-19).

FIGURE 6-18: Casting resin with metal powder mixed in (brass in this case).

Step 7

Chuckle quietly to yourself when you realize how easy it is to create convincing faux metal parts (Figure 6-20).

FIGURE 6-19: Buffing the surface to reveal the embedded metal powder

FIGURE 6-20: Hunter's belt buckle reproduced in faux brass, bronze, copper, steel, and aluminum.

Slightly Harder Version: A Piece with Details on All Sides

So you've learned how to make a mold of an object with a flat back-side, and that's fine for something that has a side that won't be visible. But what if the back-side *does* matter? Suppose we have a longer or taller prototype that has the occasional air-trapping undercut. Or suppose the prototype is shaped so that the rubber can't stretch far enough to pull the parts out of the mold—something like the huge pain-in-the-behind pictured in Figure 6-21.

No problem.

First off, it's a good idea to start by staring at the part for a while. Imagine what the mold will look like. Now imagine that mold filling up with a liquid. As the bubbles rise, is there a particular place where air will likely be trapped? Is there possibly a better place for the liquid to be poured in? All of these factors must be taken into consideration when deciding how to position the prototype for molding.

For a piece like this, the mold-making process is essentially the same as the flat-backed belt buckle, but with a few key changes. Just like the last time, the mold will be made with the bottom up and used with the bottom down. It's a good idea to keep this in mind when building this type of mold. A lot of your decisions will require you to remember that it's being made upside-down.

Here's how it goes:

Step 1

Mount the prototype to a piece of scrap wood or a heavy plastic sheet (Figure 6-22). In this case, the part isn't sitting flat. Instead, it's mounted onto a nail or screw in order to hold it upright in the mold box.

FIGURE 6-21: The Wolf Warrior Woman's dagger: a huge pain in the behind

FIGURE 6-22: The mounted prototype

Step 2

Since the silicone won't stretch enough to just slide the piece out of the mold, this mold will end up needing to be cut in order to remove the prototype. Find a good place now to make this cut later. Ideally, it'll be someplace with as few details as possible that will allow the part to be removed with the minimum amount of cutting the mold. In this case, the best place will be the narrow sides of the handle in order to free up the hand guard to pull it out of the mold (Figure 6-23). Mark this area with a permanent marker.

Step 3

Using a bit of clay, build up the area around the screw or nail into an inverted V-shape (Figure 6-24). When the mold is completed, this will become a sort of funnel (called a sprue) that will allow the resin to be poured in. A larger sprue will allow more resin to be poured in faster, but bear in mind that it will still have to be cut off and sanded smooth after the fact.

FIGURE 6-23: Marking the sides of the prototype and the bottom of the mold box where the mold will be cut

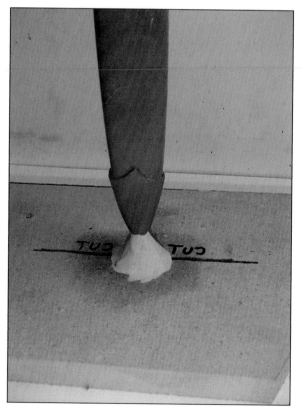

FIGURE 6-24: A sprue made out of clay

Step 4

Identify any places where bubbles flowing up toward the sprue might end up getting trapped. Remember, at this point everything's upside down, so you're looking for low points that won't allow trapped air to escape. These areas will need *vents*, or tiny channels in the rubber that will allow the air to flow up and out when the resin is poured into the sprue. At this stage, the vents can be made out of anything that won't bond to the silicone. It can be scraps of wire or plastic rod or even another nail. Just make sure it's something relatively smooth. In Figure 6-25, the vents are made out of lengths of easily cut and bendable aluminum armature wire that are glued into place.

Step 5

Build a mold box around the prototype (Figure 6-26). This part of the process is very similar to making the mold for the belt buckle. The main difference here is that the silicone, being flexible, may warp or sag under its own weight if it's too tall. This requires the mold box to be reusable. Cardboard can still work for smaller parts, but if the piece is going to be bigger, foam core, fiberboard or smooth, sealed plywood might be a good idea. Just like in the previous example, it still needs to be watertight in order to keep the silicone from oozing all over the place.

> ## 🔫 Maker Note
>
> The mold box needs to be at least ½" (approximately 1 cm) taller than the part to be molded. This helps to ensure that any trapped bubbles that don't rise all the way up to the top before the silicone cures will at least not be too close to the part being cast. For exceptionally tall parts, even more extra height is advised.

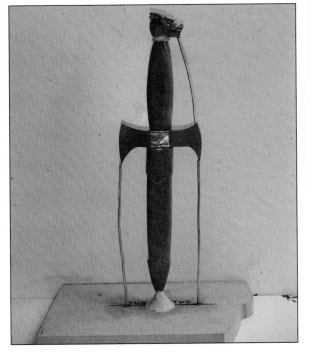

FIGURE 6-25: Vents added to allow bubbles to flow out of the mold

FIGURE 6-26: A better mold box, built to be reused if needed

Step 6

Once the mold box is watertight, mix up a batch of silicone.

Step 7

Pour (Figure 6-27). Once again, the enemy here is the air bubble. Pour the silicone into a corner of the mold as far away from the part as possible. It may also be a good idea to tip the mold box a little bit in order to prevent the rubber from being poured directly onto the top of the part and trapping bubbles underneath.

Step 8

Set the mold box upright and leave it the hell alone (Figure 6-28). Don't mess with it for at least as long as the manufacturer says it will take to cure.

FIGURE 6-28: No matter how much the mold might look like liquid candy, leave it alone.

Step 9

Once the rubber has cured, carefully remove the mold box (Figure 6-29). The main concern will be keeping the sides intact, since they'll need to be used again later.

Step 10

Make a relief cut in the mold to remove the prototype. This is more than simply hacking away with a knife. In this case, you'll be making what's often referred to as a *jeweler's cut*, designed to make sure

FIGURE 6-27: Pouring rubber with the mold box tipped slightly to allow bubbles to rise up and away from the prototype

FIGURE 6-29: Careful disassembly of the mold box

that the two sides of the cut rubber will fit back together. Begin by finding the line you originally drew on the bottom of the mold box (Figure 6-30). You'll find that it has transferred to the silicone.

FIGURE 6-30: The ink that was on the inside of the box is now on the outside of the rubber. Sorcery!

Using a hook-shaped blade, start an incision at the edge of the sprue. Make sure to cut in a serpentine pattern (Figure 6-31) with plenty of curves.

Use one hand to pry the two sides of the mold apart. Then make another cut in the bottom of the original cut (Figure 6-32). Make sure that this cut is also a curving, zig-zag shape.

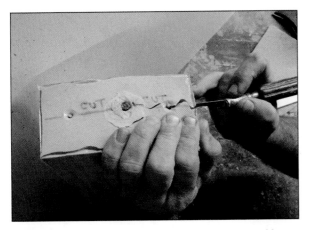

FIGURE 6-31: An initial serpentine cut in the top of the mold

FIGURE 6-32: Cutting a bit deeper

Continue to pry apart the two sides of the mold and cut deeper. It may be helpful to alternate making cuts on either side of the prototype in order to make it easier to hold the two halves apart (Figure 6-33).

FIGURE 6-33: Cutting both sides will make it easier to pry the mold open.

The aim is to make the two cut faces bumpy and uneven so that they'll fit snugly together without shifting (Figure 6-34). This way the mold won't leak when the casting material is poured in later. It'll also keep the two sides aligned so that the cast part won't be warped or misshapen.

Step 11

Demold the prototype (Figure 6-35).

Step 12

Reassemble the mold (Figure 6-36). If the relief cut was made correctly, the pieces should fit snugly together and the seam should be barely visible. Then replace the pieces of the mold box around it.

FIGURE 6-35: The prototype removed from the cut mold.

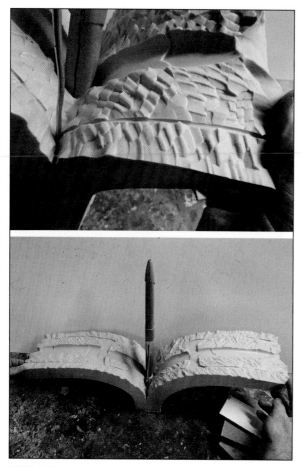

FIGURE 6-34: The jeweler's cut

FIGURE 6-36: The reassembled mold, complete with the sides of the mold box

Step 13

Mix a batch of casting material in accordance with the manufacturer's instructions.

Step 14

Pour the casting material into the sprue (Figure 6-37). At this point, it's vital to minimize the introduction of bubbles into the mold. Tipping the mold a bit and pouring slowly will keep the casting material from splashing its way to the bottom of the mold and making more bubbles.

Step 15

Jiggle it. Shaking or vibrating the mold slightly will help any trapped air bubbles rise to the surface and out of the cast part.

Step 16

Leave it the hell alone. Seriously. Enough.

Step 17

After the cast part has cured completely, demold it (Figure 6-38).

Step 18

Repeat steps 11-17 as needed until you have all of the parts your little heart desires (Figure 6-39).

FIGURE 6-37: Pouring slowly and carefully in order to avoid bubbles

FIGURE 6-38: The prototype, the mold, and the first cast part

FIGURE 6-39: A whole lot of parts

MULTI-PIECE MOLDS
When You Need to See *Both* Sides

IN THE LAST CHAPTER, you learned how to make molds out of a single block of silicone. Of course, even with vents to let the bubbles out and relief cuts to make demolding easier, one-piece molds still have their limitations. For example, suppose you're working on a piece that is a bit too long or tall to be reliably mounted inside of a mold box. There's no reason to run the risk of having the prototype fall over after pouring rubber on it, which you'll only find out after the whole thing has cured. Also, there are plenty of cases where the prototype will have details that will trap air bubbles on the sides if it's molded vertically, but this problem would be eliminated if the mold rubber was poured onto a horizontal surface instead.

While there are countless ways to approach any mold-making problem, this chapter will cover two main methods for making a multi-piece mold. First, we'll look at building a mold box and pouring rubber into it from two sides. Second, we'll discuss building up a thin rubber jacket and then laying up a form-fitted mother mold.

Starting Simple: A Two-Part Mold with a Mold Box

Take a look at the buttstock for Hunter's rifle (Figure 7-1). Those little notched details on either side would trap all sorts of bubbles in the silicone if it were to be molded while standing on end. So, instead of making a big block mold and cutting the prototype out of the block of rubber, it'll be a lot easier, and use a lot less of your expensive silicone rubber, if you make the mold in two pieces.

Tools and materials needed:

- The prototype
- Silicone RTV mold-making rubber
- Oil-based clay
- Medium-density fiberboard (MDF) or plywood to build a mold box
- Screws (or hot glue)
- An electric drill for driving screws (or a hot-glue gun for applying hot glue)
- Mold release
- Rubber bands or other suitable mold straps

Step 1

Prepare the master for molding (Figure 7-2). Clean off any loose dust or dirt (or rust or blood or bone chips). Fill in any seams or cracks that will cause the silicone to catch and rip later on. A smoother surface on the prototype will cause less friction on the silicone. This will help the mold last longer. If it can be painted with a high-gloss finish, all the better.

Step 2

Build a mold box (Figure 7-3). If you're the type of person who heats their home by burning cash, the box can be a simple rectangle large enough to hold the piece with at least ½" (1 cm) of clearance all the way around. For the rest of us, to minimize the amount of pricey silicone mold-making rubber, it's better to make a custom box. Unlike the boxes made for one-sided molds, this box will need a top as well as a bottom.

In this case, the mold box was made of scraps of MDF and plywood that were on hand. They're both inexpensive and readily available at your local home improvement store. The pieces were cut on a table saw. The box is sized and shaped to allow for at least ½" (1 cm) of clearance on all sides of the

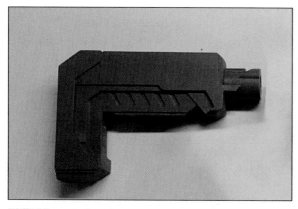

FIGURE 7-1: An ideal candidate for a two-piece mold

FIGURE 7-2: The shiny, clean prototype all ready to mold

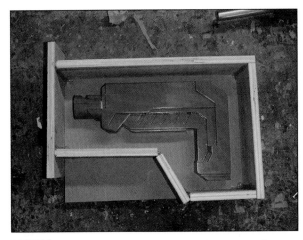

FIGURE 7-3: A mold box made to fit the prototype just right

FIGURE 7-4: The clay bed built up for the prototype to rest on.

original piece. This includes making it tall enough for the prototype to have ½″ (1 cm) clearance above *and* below it.

touches the prototype) should be as close to horizontal as possible.

> ## 🔫 Maker Note
>
> When building the mold box, fasten it together temporarily with screws or hot glue so you can take it apart later. If you make the seams permanent at this stage, you're going to regret it. Maybe not today, and maybe not tomorrow, but soon, and for the rest of your life.

> ## 🔫 Maker Note
>
> Make sure that all of the edges of the clay wall are watertight. The last thing you want is to have silicone drooling out all over your workspace. It's considered a major mold-making faux pas.

Step 3

Build a clay bed. Start by laying down a few blocks of clay at least ½″ (1 cm) thick to set the prototype onto (Figure 7-4). If the mold box is tall enough, it can be thicker. Just remember that you'll still need ½″ of clearance above the highest part of the prototype.

Once the prototype is securely held in place, build up more clay in the areas between the prototype and the mold box (Figure 7-5). The top of the clay will need to be smooth and the edges (where it

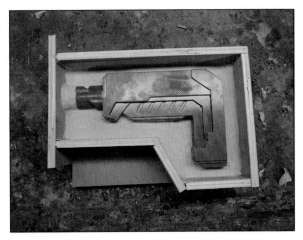

FIGURE 7-5: Filling in the area around the prototype with more clay

Step 4

Add in sprues and vents, as needed (Figure 7-6). After determining where the rubber will be poured in, vents can be made by pressing wire lightly into the clay. If you make a mistake at this stage, more vents can be added by cutting grooves in the rubber later—but it'll be a bit more difficult at that stage.

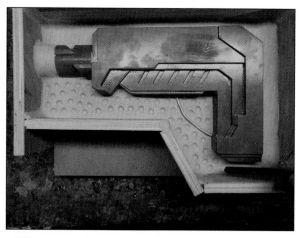

FIGURE 7-7: The finished clay bed, complete with dimples to form registration keys

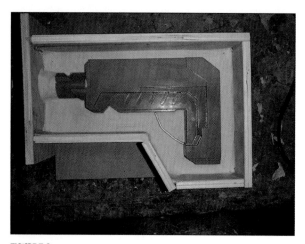

FIGURE 7-6: Sprues will allow the casting material in, while vents will let the air out.

Step 5

Add registration keys to the clay bed (Figure 7-7). The clay bed will determine the shape of the second half of the silicone mold, so you'll be designing the *parting line* where the two halves of the mold will fit together. Using the round end of a sculpting tool, push a bunch of dimples into the clay wall, taking care not to open up any edges where silicone will leak out. The last thing you'll need to do is to make *registration keys*. These are dimples that will help to keep the two halves of the mold properly aligned when it comes time to make castings.

Step 6

Enter the goop! After mixing the rubber in accordance with the manufacturer's instructions, it's time to pour it into the mold (Figure 7-8). At this point, you want to avoid introducing any additional air bubbles into the mixture. Bubbles will show up as flaws on the cast pieces, and those flaws will take time to repair. Once again, time saved by cutting corners at this stage will be paid for later—with interest.

As with the one-piece mold, start by pouring the rubber into a corner and allowing the material to flow over the part(s) in the mold box.

FIGURE 7-8: Pouring rubber with style

playing the ukulele. Maybe go outside, or shopping . . . or finally ask that someone special out on a date. In other words, do whatever you need to do to keep yourself from poking your sticky boogerhooks into the freshly poured rubber before it cures.

🔫 Maker Tip

After you've been doing this sort of thing for a while, you'll eventually have a few old molds sitting around that you don't need. Hang on to them. Cut-up chunks of old, worn-out molds can be sunk into new the batches of rubber as new molds are being poured. This sort of "econo-fill" is a great way to recycle your dead molds and save on expensive silicone rubber. Just be careful to avoid capturing bubbles under the solid chunks of rubber.

🔫 Maker Note

To impress your friends, raise the mixing container to seemingly ridiculous heights while pouring. This isn't just for panache (although passers-by can't help but be impressed with such rubber-pouring flair). The rubber stretches out to a tiny thread as it falls, popping any entrapped air bubbles on the way down.

Step 7

Once you've filled the mold box all the way up to the tippy top, leave the mold to cure for however long the manufacturer suggests. Go home and get a good night's sleep. Or read a book. Practice

Step 8

Attach the top onto the mold box. Then, flip the whole thing over and remove the bottom (now the top) of the mold box (Figure 7-9). This is why you'll be glad you used screws or hot glue to hold the mold box together.

FIGURE 7-9: The mold box flipped over with its bottom removed

Step 9

Remove the clay (Figure 7-10). As you can see, everywhere there was a dimple in the clay there is now a matching bump in the newly cured rubber; this will become an important plot element as this story unfolds. At this point, the only clay that should still be in the mold box is the little bits left in the corners to prevent it from leaking.

FIGURE 7-10: Bumpy silicone rubber

Step 10

COAT THE WHOLE THING WITH RELEASE AGENT! If you only remember one thing about silicone mold-making rubber, it's this: silicone sticks to almost nothing except for more silicone. That, and don't eat it. Two things. Dang it.

Before you pour the second half of the mold, something must be done to ensure that the two halves won't bond together. This calls for some kind of release agent. Petroleum jelly will work, but it's messy. The better option is to spray on an extremely thin coat of whatever the silicone manufacturer recommends. This will keep the two halves from bonding, but being thin means that they will still fit snugly together without any casting material leaking out.

Some folks might tell you that a little will go a long way. They'll also suggest that a lot does no better. Still, if in doubt, use a bit more. Then maybe even a bit more than that. In any case, read the instructions. It's the only way to be sure.

If you forget or choose to skip this step, all of that time spent sculpting that beautiful clay bed will have been for naught.

Step 11

Pour the second half of the mold (Figure 7-11). Pour it just like the first half.

When the box is filled all the way to the tippy top, it's time to set it on a level surface to cure. Once again, don't mess with it until it's fully cured.

Step 12

Take the sides off the mold box (Figure 7-12).

FIGURE 7-11: Pouring the second half just like the first half

FIGURE 7-12: The disassembled mold box

Step 13

Gently peel the two halves of the mold apart and pull the prototype out of the mold (Figure 7-13). If the prototype breaks in the process, worry not. It's about to be replaced.

Step 14

Put the mold back together, leaving off the top and the side with the sprue (Figure 7-14). This time, the sides of the mold box can be glued together,

making it permanent. It will not have to be completely disassembled again. This is in accordance with the prophecy.

🔫 Maker Tip

Before putting the mold box back together, it's often a good idea to take the side pieces back to the table saw and shave a bit off them to make the mold a little shorter. This way, when the lid is bolted down, it will serve to clamp the two sides of the rubber together and prevent leakage. Don't bolt it too tightly, though, because doing so will squeeze the mold and distort it, making the castings misshapen.

Step 15

Attach the lid, as shown in Figure 7-15. For a smallish mold like this, rubber bands will be able to apply more than enough pressure to keep the casting resin from leaking out between the two halves. If the mold were larger, bar clamps might be a good idea.

FIGURE 7-13: The demolded prototype

FIGURE 7-14: The reassembled mold box, without top

FIGURE 7-15: The mold box strapped together with rubber bands

Step 16

Mix a batch of casting resin and pour it into the mold (Figure 7-16). Once again, it's vital to avoid the introduction of any air bubbles. Tipping the mold a little bit to one side will allow the resin to flow down the side of the mold instead of just falling to the bottom and splashing on impact.

FIGURE 7-16: Pouring in some resin

Step 17

After the resin has fully cured, remove the lid and pull the cast part from the mold (Figure 7-17).

🔫 Maker Note

For larger parts cast in urethane resin, it is often a good idea to add filler known as microballoons or microspheres. These are tiny, hollow, glass bubbles that can significantly reduce the weight of the final part. Doing this will also cut down on the total volume of expensive casting resin needed to fill the mold. The trade-off is that it will cause the finished part to be a bit less sturdy. Ask your resin supplier for details.

FIGURE 7-17: The first cast part pulled out of the mold

The masterpiece is complete. Bask in its glory and wonder. Stand up. Applaud. Clasp it to your breast, weeping in celebration, as men do.

Step 18

Repeat steps 15–17 until you have all of the copies you could possibly want (Figure 7-18). Made properly, a mold like this can potentially make dozens and dozens (and more dozens) of castings, before it wears out.

But what if the part is even bigger? Unless you're willing to sell a kidney on the black market to help finance your silicone mold-making needs, you'll need a more efficient way to use the rubber than just filling up big mold boxes.

Not So Simple: A Three-Piece Silicone Rubber Jacket Mold with a Mother Mold

While the two-part box-mold method makes sense in a lot of cases, at some point you'll be molding a piece that's so big that it'll be too expensive to just make a big block of rubber. In that case, it makes more sense to build the rubber in a thick, form-fitting "jacket" that can make castings that are just as good as a big block mold, but don't use nearly as much rubber.

Since the rubber will warp or sag under its own weight, however, you'll also need to build a rigid shell that the jacket will fit into. This will keep everything straight during the casting process. It sounds complicated, but it's only slightly more difficult than the molds that've been explained so far.

FIGURE 7-18: Parts piling up

What's more, some pieces require the silicone to be poured on from more than two sides to avoid trapping bubbles against the surface. Looking at the main body of Hunter's rifle (Figure 7-19), it could almost be split down the middle and poured from the left and right sides. The problem is, the magazine well on the bottom would probably trap air bubbles if you did that.

To get around this problem, we'll need to make a three-part silicone rubber jacket mold with a mother mold. Start by getting your materials together.

FIGURE 7-19: The magazine well on the bottom of Hunter's rifle adds a bit of a challenge to the mold-making.

Tools and materials needed:

- The prototype
- Silicone RTV mold-making rubber
- Thixotropic additive for the silicone rubber
- Oil-based clay
- Sculpting tools—nothing fancy; just something that can smooth the clay
- Fiberglass mat
- Fiberglass resin
- Electric drill
- ¼" drill bit
- Mold release
- ¼-20 x ¾" hex bolts
- ¼" wing nuts
- ¼" washers
- Disposable bristle brushes
- An ice cube tray that will never be used for ice again

Once all of the requisite tools and materials are rounded up, it's time to come up with a plan.

Take a look at the prototype. Imagine a hollow glass container of the same shape with no holes. If you had to fill it with water, what would be the best place to drill a single hole that would allow you to pour in water without any air getting trapped inside?

In the case of Hunter's rifle body, it looks like it will work fairly well if we stand it up on end, barrels pointing upward, and fill it from the front end. As luck would have it, once the mold is made, we'll be able to pour resin in through one of the barrel tubes while the air flows out through the other tube.

Now that we have a plan, it's time to get stepping.

Step 1

Prepare the prototype (Figure 7-20). It's still going to be a piece that will be molded with silicone rubber, so it still needs to be as clean and smooth as possible. It's also a good idea to draw a line down the center of the prototype where the two halves of the mold will separate. While it's usually not necessary to use a release agent to keep silicone mold-making rubber from sticking to the prototype, it will increase the life of the mold, while eliminating any chance that the rubber will bond to the prototype.

FIGURE 7-20: The prototype all ready for molding

Step 2

Set up your work space. You'll want a stable working surface that's free of obstructions. You'll also want a layer of newspaper or some other disposable covering to keep a spill from causing you to be yelled at by whoever else uses the space. This applies even if the only other person using the workspace will be the future version of you who will rue the day you dribbled molding and casting materials all over the formerly pristine kitchen table or living room floor.

🔫 Maker Tip

For maximum versatility, it's a good idea to put a piece of MDF or plywood down first as a base for the mold-making. This way, you'll be able to move the work-in-progress to a shelf or other out-of-the-way area while waiting for it to cure.

Step 3

Stabilize the prototype and build a clay wall around it. Just like the two-part box mold, the prototype will need to be mounted into a bed of clay. Unlike the two-part box mold, it doesn't need to be raised off the work surface at all. Instead, scraps of whatever's handy can be wedged around it to keep it from wobbling around while the clay bed is built (Figure 7-21).

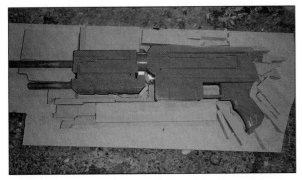

FIGURE 7-21: The prototype stabilized with pieces of whatever's handy

🔫 Maker Note

Unless you want to build up a massively heavy pile of clay that's over half as thick as your prototype, you can get away with using scraps of wood to fill up most of the space around the prototype. You can use bits of clay stuck to the working surface to hold these scraps in place and keep everything from slipping around while you work (Figure 7-22).

FIGURE 7-22: Blobs of clay to stick the blocks of scrap to the work surface

Once the prototype is wobble-free, build up the clay bed around it (Figure 7-23). Take special care to ensure that the edges where the clay meets up with the prototype are watertight so the liquid silicone rubber won't leak all over the place. The clay bed should have at least 2″ (5 cm) of margin around the entire prototype.

FIGURE 7-23: The clay bed with a good margin around the prototype

> ### 🔫 Maker Note
>
> At the bottom of the rifle is the magazine well. That's the hole where the ammunition magazine gets inserted. To minimize the chances that it will trap bubbles and cause problems during the rubber-pouring stage, that area will be molded separately. So at this stage, it's important to build up the clay wall to cover the entire inside of the magazine well, as shown in Figure 7-24.

Using a rounded object, such as a dowel or the back end of a sculpting tool, press dimples into the clay around the part, as shown in Figure 7-25.

Take a close look at the edges where the clay meets up with the prototype (Figure 7-26). Be sure to remove any residual clay that remains on the prototype and verify that the edge is watertight all the way around the prototype.

Since there's no mold box to contain the rubber, build a drip wall around the outside edge that's at least 1" (2 cm) tall on top of the clay bed and about 1" (2 cm) away from the prototype (Figure 7-27).

FIGURE 7-25: The dimpled clay bed

FIGURE 7-24: The magazine well filled with clay

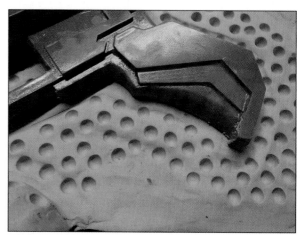

FIGURE 7-26: Any tiny clay remnants left on the prototype at this stage will become flaws on the castings later.

FIGURE 7-27: The clay bed, complete with a raised drip wall around the outside

Maker Note

The clay bed extends outward past the drip wall. This excess will form the flanges of the mother mold in a later step. It does need to be there.

Step 4

Triple-check to make absolutely, positively certain, beyond a shadow of a doubt, that the clay bed and drip wall are watertight. The last thing you want is to find out you left a pinhole somewhere and have a bunch of expensive silicone drool all over the table and floor.

Step 5

Pour a *print coat*. Also referred to as the *detail coat* or *beauty coat*, this first coat of silicone is going to be the one that will seep into all of the details. Mix only enough to completely cover the prototype and the bottom of the clay bed. When pouring the print coat, be sure to cover every bit of the prototype (Figure 7-28). This will mean pouring the liquid rubber onto the high parts as well as the low parts, so extra care must be taken to ensure that no air bubbles are trapped under the rubber.

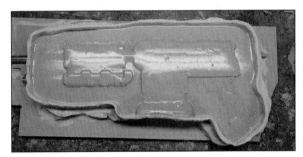

FIGURE 7-28: The print coat

Registration Keys

When you pour the print coat, it's also a good idea to pour a batch of registration keys. These will be noticeable lumps of silicone that help align the rubber jacket mold inside the mother mold. Specialized molds for silicone registration keys can be purchased at the same places that sell silicone, but an ice cube tray can work just as well (Figure 7-29).

Just take care not to make fill the ice cube tray all the way to the top. Halfway should be fine. If the registration keys are too tall, they can cause a lot of problems.

FIGURE 7-29: The half-filled ice cube tray makes good registration keys.

Maker Note

If a vacuum chamber is available, degassing the silicone before pouring the print coat is a very good idea.

Step 6

After the print coat has cured, layer on thickening coats. There are two options for this:

OPTION 1   Mix a batch of silicone as per the manufacturer's instructions. Pour it onto the prototype just like the print coat. Then, after it drools all over the place and settles in the lowest points of the prototype and the clay bed, scoop it out of the lower areas with a spoon or mixing stick and ladle it over the high points again. Eventually, the silicone will begin to thicken or *gel* and become more viscous.

The object of the game is to repeat this process, mixing, pouring, and ladling more and more silicone until at least ½" (1 cm) of rubber has built up over the entire prototype, as well as on the bottom of the clay bed. This will take a lot of time, but with care it will result in nice, smooth molds with a minimum of air bubbles.

OPTION 2   Mix a batch of silicone in accordance with the manufacturer's instructions. Then mix in a thixotropic additive, which should be available from the same supplier as the silicone itself. A *thixotropic additive* is a chemical that will cause an increase in the viscosity of the mixed silicone. Exactly how thick the mix will be can usually be adjusted by changing the amount of additive that is added into the silicone.

Once the silicone is mixed, it can be spread on top of the print coat much like frosting a cake (Figure 7-30). Just remember that, although it usually ends up looking like candy, you must not taste the silicone.

Step 7

Once the silicone is built up to be at least ½" (1 cm) thick over every part of the prototype, add registration keys by sticking them into the still-sticky surface of the silicone (Figure 7-31). There's no hard and fast rule regarding how many you'll need or how far apart they should be. Just make sure there are enough of them so that there's only one possible way to fit the rubber jacket into the rigid mother mold.

Step 8

Take a serious, judgmental look at it. Is the surface nice and smooth? Chances are it's not. Whichever method was used to thicken the rubber, there will

FIGURE 7-30: The rubber jacket halfway frosted with thickened silicone

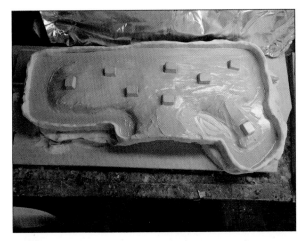

FIGURE 7-31: Registration keys added to the surface of the mold

FIGURE 7-32: Smoother mold jacket

probably be at least a few ridges or spikes sticking out of it. This is usually because you just couldn't leave well enough alone and had to keep messing with it, even though the silicone had begun to gel and was well past its working time. You should be ashamed of yourself.

Having all of these extra nooks, crannies, ridges, valleys, protrusions, depressions, bumps, and holes will make it harder to get the mother mold to fit properly onto the rubber jacket. In order to smooth it all out, pour on one more layer of silicone and let it settle (Figure 7-32). Don't add any thixotropic additive. Don't ladle it from the low points to the high points. Just leave it alone until it has cured.

FIGURE 7-33: His name was Mortimer, the mosquito hawk, and he carried out his duties tirelessly, with diligence and care—except for that one time.

Step 9

Leave it alone until the silicone has cured completely. Seriously. It's a good idea to cover it with something to keep suicidal critters from dying in the rubber (Figure 7-33).

Step 10

Remove the clay drip wall. Make the clay underneath flat and smooth (Figure 7-34).

FIGURE 7-34: The drip wall removed

Step 11

If there are any ridges or high points along the edge where drips have cured against the drip wall, remove them with a sharp hobby knife (Figure 7-35).

FIGURE 7-35: Cleaning up the edge of the rubber

Step 12

Build the first half of the mother mold. While it is possible to make the mother mold with plaster bandages or a variety of proprietary materials that are designed especially for the purpose, one of the best options available is a quick fiberglass layup. The materials can be found at the local hardware or auto parts store, and the end result will be strong and lightweight.

🔫 Maker Note

Fiberglass, also known as *glass reinforced plastic*, is an example of what's called a *composite material*. This just means that it's composed of more than one material. By itself, the resin is brittle and easily broken. The tiny glass strands are flexible but difficult to break. Separately they're pretty useless, but, by combining the two, the resulting composite gains the advantages of both materials.

Start by mixing up a batch of fiberglass resin, and then use a disposable bristle brush to paint a layer onto the silicone and the clay bed (Figure 7-36).

FIGURE 7-36: A coat of fiberglass resin painted onto the silicone and the clay bed

Tear off hand-sized patches of the fiberglass mat and place them on top of the still-wet coat of resin (Figure 7-37). Make sure the edges of the patches overlap slightly and that the mat covers everything, all the way out to the edges of the clay bed.

FIGURE 7-37: Fiberglass mat laid evenly over the wet resin

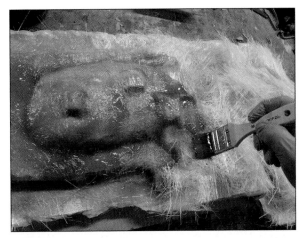

FIGURE 7-38: Brushing resin into (and bubbles out of) the fiberglass mat

Use the brush to apply more resin on top of the fiberglass mat (Figure 7-38). Add only enough resin to saturate the mat all the way through. Along the way, use the brush to force out any bubbles that are trapped in the mat.

Once the first layer is completely saturated (often called *wetting out*), tear off more patches of fiberglass mat and build up another layer just like the first. Take care to work out all of the bubbles along the way. Since the resin in the first layer is still liquid, it will take less resin to wet out the second layer.

Repeat this process until there are at least three layers of fiberglass mat and resin composite built up over the entire piece (Figure 7-39). This should be adequate for a nice, strong mother mold. If in doubt, more can be added, but it shouldn't be necessary for a small piece like this.

Step 13

Leave it to cure completely. Depending on the ambient temperature and the amount of catalyst mixed into the fiberglass resin, this can take anywhere from 45 minutes to several hours. Don't mess with it!

FIGURE 7-39: The first half of the mother mold completed

Step 14

Flip the whole thing over and remove the scraps of wood and wedges that were used to keep everything in place (Figure 7-40).

Step 15

Remove the clay from the inside area of the clay bed, leaving a border in place around the perimeter of the mold. Notice that every dimple that was pushed into the clay has now caused a bump to form in the silicone. Nifty!

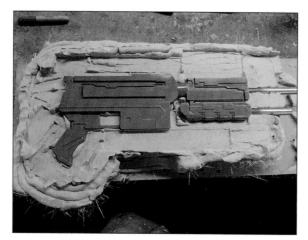

FIGURE 7-40: The underside of the clay bed is kind of a mess.

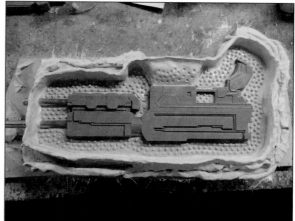

FIGURE 7-41: New drip walls built up on the mother mold

Build new clay drip walls on top of the fiberglass mother mold (Figure 7-41). Just like before, take care to ensure that they're watertight.

Step 16

SPRAY ON THE MOLD RELEASE! This step is absolutely vital. Mold release is transparent, so a photo wouldn't mean much. Just don't forget to put it there.

Step 17

Mix a batch of silicone in accordance with the manufacturer's instructions. Then, pour on the print coat (Figure 7-42). Once again, the object is to have a nice, bubble-free layer of silicone flow over the part so that all of the details will be replicated.

Step 18

Once the print coat has cured, build up thickness just as you did with the first half (Figure 7-43).

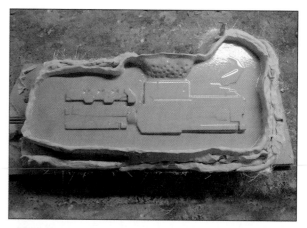

FIGURE 7-42: The print coat poured on the second half of the prototype

FIGURE 7-43: The thickened rubber jacket

Step 19

Add registration keys, as needed (Figure 7-44).

Step 20

If there are any hooks, spikes, or ridges on the surface of the silicone jacket, pour on one more coat of unthickened silicone (Figure 7-45). This will fill up any nooks and crannies, and settle into the low areas.

Step 21

Wait for the silicone to cure. You don't have to sit there and watch it. Just cover it with something to keep stray bugs or dust from landing in it and go do something else. Consider playing with a puppy (Figure 7-46).

Step 22

Once the silicone has completely cured, remove the clay drip walls (Figure 7-47).

Step 23

Add a few pieces of clay along the edges of the fiberglass mother mold (Figure 7-48). These will form gaps between the two halves of the mother mold. This will make "prying points" where a tool can be inserted to force the two halves of the mold apart once the whole

FIGURE 7-44: Déjà vu is setting in.

FIGURE 7-46: A puppy is always great for passing the time.

FIGURE 7-45: One more smooth coat of silicone

FIGURE 7-47: At this point, all of the clay has been removed except for the plug in the magazine well.

thing is laid up. They should be just wide and thick enough to fit a screwdriver a little way into them.

Step 24

Apply more mold release (Figure 7-49). Fiberglass resin is designed to stick to everything, especially more fiberglass resin. In order to prevent the two halves of the mother mold from bonding together to form a bulletproof enclosure that will permanently encase the prototype for all eternity, coat it with something that will create a barrier layer between the parts. In a pinch, petroleum jelly works; just don't lay it on too thick.

Step 25

Build the second half of the mother mold (Figure 7-50).

FIGURE 7-48: Lumps of clay along the edges of the mother mold

FIGURE 7-49: Applying petroleum jelly as a release agent. Icky.

Step 26

Drill a series of ¼" holes around the outside edges (the flanges) of the mother mold (Figure 7-51). Make sure that they are on flat areas so that bolts can later be inserted to hold the two halves of the mother mold together.

Step 27

Make the plug for the magazine well. Start by bracing the mold so that it will sit upright with the magazine well facing upward (Figure 7-52).

FIGURE 7-50: Build the second half, same as the first half.

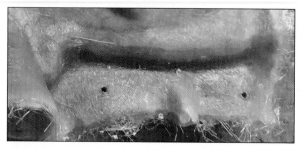

FIGURE 7-51: Bolt holes drilled in the flanges of the mother mold

FIGURE 7-52: The mold wedged into position for the last section to be made

Remove the clay from inside the magazine well and spray the recess with mold release (Figure 7-53).

Mix a batch of silicone big enough to fill the magazine well, leaving about 1" (2 cm) empty at the top, as shown in Figure 7-54.

Once the silicone plug has cured, goop some petroleum jelly onto the exposed portion of the fiberglass flange, then lay up a cap that will fit into the recess and hold the plug in place. When the glass layup has cured, it's time to trim off the rough edges, and drill a few holes around the edges so it can be bolted into place (Figure 7-55).

Step 28

Take the mother mold apart. Start by removing all of the nuts and bolts, and then insert a pair of screwdrivers, or similarly shaped tools, into the clay-filled prying holes. Now it's time to gently pry the two halves of the mother mold apart (Figure 7-56). Once the two halves start to come apart, work the tools around the entire outside edge until the halves are completely separated.

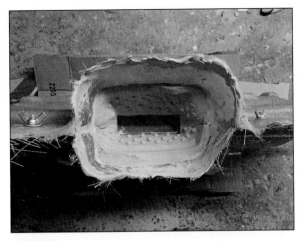

FIGURE 7-53: The magazine well prepped for molding

FIGURE 7-55: The cap over the magazine well plug all fitted with bolts

FIGURE 7-54: The hole filled with more rubber, but not completely filled

FIGURE 7-56: Prying the two halves of the mother mold apart

Step 29

Peel the rubber jacket off the prototype (Figure 7-57).

Step 30

Place the two halves of the rubber jacket into their respective mother mold halves (Figure 7-58).

Step 31

Bolt the two halves of the mother mold together (Figure 7-59), using washers to spread the strain, and using wingnuts to make life easier.

Step 32

Mix a batch of casting material and pour it into the sprue (Figure 7-60).

Step 33

Wiggle, jiggle, and vibrate the mold to help work the air bubbles out. You'll like it.

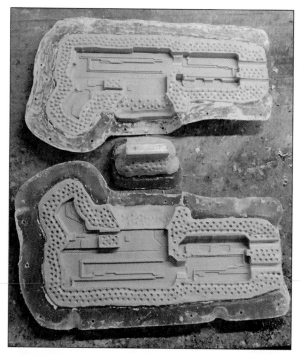

FIGURE 7-58: Rubber jacket halves fitted into mother mold halves

FIGURE 7-57: Welcome back, prototype buddy.

FIGURE 7-59: Wingnuts are the best!

Step 34

Once the cast part has had plenty of time to cure, unbolt the two halves of the mother mold and remove it (Figure 7-61).

FIGURE 7-60: Pouring a casting

FIGURE 7-61: Fresh, hot, cast part removed from the mold

Step 35

Repeat as necessary to get all of the castings your heart desires (Figure 7-62).

A Few Tips and Tricks for Better Castings

If you notice that there are tiny air bubbles on the surface of the cast parts (Figure 7-63), it may help to dust the mold lightly with talcum powder before pouring the casting. This will break the surface

FIGURE 7-62: Cast parts for all your friends

FIGURE 7-63: Tiny bubbles are the bane of a prop maker's existence, and they must be destroyed.

tension of the resin that is poured into the mold and keep bubbles from clinging to the silicone.

For some pieces, it may be a good idea to coat the two halves of the mold with resin before bolting them together (Figure 7-64). This will allow gravity to help the resin work its way into the details, and make life easier all the way around. Once the resin has begun to gel, the two halves can be bolted together and then you can pour as usual.

As previously mentioned, in the case of very large parts, microballoons (also known as *micro-spheres*) can be mixed into the casting resin to reduce the weight and cost of the final cast part. Be careful, though. The more you add, the more break-able the final piece will be.

FIGURE 7-64: Coating the two halves separately puts gravity to work for *you* for a change.

8

MOLDS FOR ROTOCASTING
Bigger Doesn't Have to Cost More

IN PREVIOUS MOLD-MAKING CHAPTERS, we've covered molds that are used to cast solid parts. But sometimes it's not such a good idea for a piece to be solid. Whether it's to cut down on weight, reduce costs, or just make it a bit easier to fit your head inside, sometimes you need to make a hollow part. It's better than the option shown in Figure 8-1!

FIGURE 8-1: Dressing in costume is more fun when you don't have to start with decapitation.

Tools and Materials Needed

The prototype

Respirator

Rubber gloves

Safety glasses

Silicone RTV mold-making rubber

Thixotropic additive for the silicone rubber

Oil-based clay

Cardboard

Glue (strong stuff)—hot glue will work great

Sculpting tools—nothing fancy, just something that can smooth the clay

Wooden dowels or plastic rods cut into 1″ (2 cm) lengths

Fiberglass mat

Fiberglass resin

Mixing cups

Stir sticks

Ice cube tray or other shallow mold for making registration keys

An electric drill for drilling holes

¼″ drill bit for drilling holes

Mold release

¼″-20 × 3/4″ hex bolts

¼″ wing nuts

¼″ washers

Disposable bristle brushes

Let's take a look at the Hunter helmet prototype that was built in Chapter 2. The main problems with the prototype were that it was fragile, heavy, and pink (Figure 8-2). To make it into something wearable, we need to keep it exactly the same size and shape, but fabricate it from a different material, and possibly make it a different color, as well.

Time to make another mold!

A mold box big enough to fit this thing would require a small fortune in silicone rubber to fill. What this project calls for instead is a rubber *jacket mold* with a rigid *mother mold*. Then, rather than simply pouring resin in until the mold is full, the inside will be coated with resin using a process called *rotocasting* or *slush casting* to make hollow copies.

FIGURE 8-2: The heavy, fragile, pink prototype

Once you've gathered up all of the tools and materials, it's time to begin.

Prepare the Prototype

Before making the mold, we need to make sure the prototype is ready to go. This means that the paint needs to be dry and the whole thing needs to be free of dust and debris. Look it over to make sure that there are no unnecessary rough spots or holes where the silicone is likely to seep in and try to lock on mechanically.

Once the prototype helmet is cleaned up, start by closing the neck hole, as shown in Figure 8-3.

When rotocasting something like a helmet, it's often difficult to see that the bottom edge is adequately coated with resin. To make sure this notoriously weak area will be sturdy enough once the part is cast, it helps to build up a lip that will catch a bit of resin in the bottom of the helmet.

Of course, since the mold is an inside-out version of the shape to be made, making a lip in the mold actually means making a groove in the prototype. In Figure 8-3, notice that the cardboard

was glued into the helmet, leaving a bit of a recess inside the neck hole.

If the glued seam between the helmet and the neck hole isn't smooth, it's a good idea to smooth over it with some clay. Try to make the inside edge as smooth as possible, but bear in mind that nobody but the wearer will ever see what this edge looks like. Make sure the clay makes a good seal all the way around between cardboard and the edge of the helmet.

With the lip around the edge of the helmet filled in, the next step is to build a cardboard stand to support the helmet. This is basically just a tube that will keep the helmet upright when it's standing on the table. This should also be glued in place, making sure that it is watertight where it's attached to the cardboard, closing off the bottom of the helmet (Figure 8-4).

Ideally, it should keep the helmet about 1" (2 cm) above the table when the helmet is sitting upright. More than that will mean wasting expensive silicone rubber. Less will leave too little room to work around the bottom edge. This will all make sense in the end.

FIGURE 8-3: Cardboard hot-glued into the bottom of the helmet

FIGURE 8-4: The cardboard wall that will later support the helmet

Finally, since the rubber jacket will have to be cut in order to peel it off the prototype and each of the cast copies, now is a good time to determine the best place to cut. Ideally, the split in the mold should be placed in an area where it will be easy to sand down any resultant *flashing* or excess resin flaps that leak through the cut in the rubber.

Once you've picked out a good place to make the cut, it's a good idea to mark the line to be cut with a permanent marker.

Now that the prototype is ready to mold, it's time to . . .

Prepare the Work Space

Ideally, you will want a work surface that's large enough to support the piece being molded, but small enough to allow you to work on it from all sides. If the just-right–sized table isn't handy, no worries. Use a scrap of plywood or cardboard as a base that you can slide around on the table or workbench so all sides of the prototype can be reached.

It's also a good idea to cover the floor and any other surfaces that might not want to be splattered with mold-making rubber. This is a potentially sticky process with a lot of icky chemicals to be used along the way.

In this case, the mold will be a single piece of rubber that will end up being cut in order to remove the prototype. Since there are significant undercuts that will trap and hold air bubbles when the rubber is poured on, it's a good idea to start with the helmet inverted. To keep it from rolling around, block it up with wedges or rolled-up bits of cardboard—or just jam it into a snug-fitting box while you're pouring.

Building Up the Rubber Jacket

Mix up a batch of silicone and fill in the recessed area between the cardboard stand and the edge of the helmet (Figure 8-5). Fill in any areas that will be hard to reach when the helmet is turned upright. Take care to avoid letting the liquid rubber drool all over and glue the prototype to the table.

FIGURE 8-5: Filling the undercuts with silicone while the prototype is upside-down

Let it sit upside-down until the rubber has cured on the bottom. Don't rush it. The anticipation may be overwhelming, and it's often tempting to start messing with the rubber to try to rush the process, but patience will always pay off in projects like this.

Once the silicone has cured, flip the whole thing right side up and mount it to a piece of cardboard with hot glue, as shown in Figure 8-6. Make sure that the cardboard base is at least 1" (about 2 cm) wider and longer than the piece being molded.

Now, glue down a series of pegs on the base around the outside of the helmet (Figure 8-7).

FIGURE 8-6: The helmet stand glued to a base

The last thing to do before pouring on the rubber is to cut a few 1"-wide (about 2 cm) strips of cardboard that can be used as a "drip wall" around the bottom of the prototype, as shown in Figure 8-8. This will keep the silicone from dripping off the part and flowing all over the table, the floor, and everything in between. The drip wall will need to be glued down watertight in order to work.

Now that the prototype is properly mounted, it's time to mix up a batch of silicone and pour a

> ### 🔫 Maker Note
>
> The drip wall should be at least 1" (about 2 cm) away from the area directly under the prototype. This way, anything that drips off will still stay contained.

set of registration keys in a shallow mold, such as an ice cube tray, or a purpose-made mold that can be bought at some art supply stores. You'll want the keys to be ¼" to ½" thick (6 mm–10 mm). Any thicker would be a waste of expensive silicone and would make it more difficult to build the mother mold to fit them properly.

Next, mix another batch of silicone and pour a thin coat over the whole thing (Figure 8-9). This first coat, often called the *print coat*, is the one that has to pick up all of the fine details of the surface. Allow it to flow over the entire surface of the prototype. When the silicone flows to the bottom, scoop it up with a spoon or a stir stick and pour it back onto the top of the prototype.

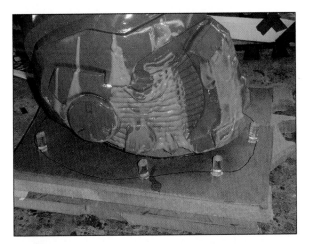

FIGURE 8-7: These bits of rod (or dowel) will become alignment pegs later.

FIGURE 8-8: The cardboard drip wall

Along the way you'll notice that some of the deeper recesses will form bubbles (Figure 8-10). Keep an eye out for these areas and be sure to pop them with your mixing stick. Ensure that plenty of bubble-free rubber is forced into the area instead.

Eventually, the silicone will gel up and stop flowing. At this point, you should stop messing with it and allow it to cure. Be sure to cover the whole thing with a box or bucket or some such thing in order to keep dust and bugs from sticking to the wet rubber, as shown in Figure 8-11.

When the rubber has cured, there will be places where the silicone is very thin and the high edges are showing through (Figure 8-12).

To make sure that the mold is thick enough to resist tearing when the castings are pulled, you need to add more silicone. This is where *thixotropic* additive comes in. Most silicones are going to flow freely to allow bubbles to work their way up and away from the part being molded. Mixing in a thixotropic additive makes it viscous enough to cling to vertical surfaces, as shown in Figure 8-13.

FIGURE 8-9: Pouring begins

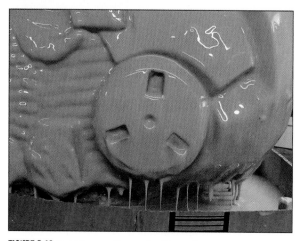

FIGURE 8-10: Bubbles are the work of Satan!

FIGURE 8-11: To the rest of the world Steve was just a fly—a nuisance, easily forgotten. But to me he was a friend. So it goes.

FIGURE 8-12: The cured print coat is thinner in high areas where the liquid rubber flowed away.

FIGURE 8-13: Thixotropic additive makes this stage a bit like frosting a cake.

The object is to make the silicone jacket thick enough to hold up to the strains of pulling out the castings and to keep its shape without becoming distorted in the casting process. But too much silicone will make the whole thing uncomfortably heavy during rotocasting (not to mention a waste of expensive silicone rubber). You will probably have to work through a couple of batches of silicone to get it right.

> ### 🔫 Maker Note
>
> Unlike the usual mixture, the thickened silicone will not allow bubbles to rise to the surface and flow out. If you have access to a vacuum chamber, it's a good idea to use it to evacuate any trapped air from the rubber. If not, take extra care when spreading the rubber over the surface to force the air out along the way.

Ideally, the silicone jacket should be built up until it's at least ½" (1 cm) thick over the entire prototype, with an even thicker ridge built up where

it will be cut open. The surface should also be smoothed out as much as possible. Little ridges and spikes left behind when troweling on the rubber will make it harder for the rubber jacket to fit into the mother mold later.

> ### 🔫 Maker Note
>
> At this point it's often a good idea to add a bit more thickness along the edge where the rubber will be cut later.

While this last layer of silicone is still tacky, it's time to add the registration keys that were poured at the beginning of the silicone jacket building stage (Figure 8-14). These will provide obvious points where the jacket will fit into the mother mold in order to make sure everything is properly aligned and avoid warping the rubber jacket during the casting process. Just remember that they'll have to be positioned so that they won't be ripped off by the mother mold when the rigid shell is removed.

FIGURE 8-14: Silicone registration keys stuck into the tacky silicone surface.

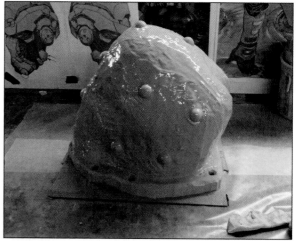

FIGURE 8-15: Pegs and drip wall removed from the bottom of the jacket mold

Once the mold is thick enough and the registration keys have been added, take a good hard look at it. Is the surface of the rubber smooth, or are there little bumps and ridges and holes all over it? If it's not smooth, mix up another small batch of silicone and pour it over the whole thing. This helps eliminate all the little barbs and snags that would lock into the rigid mother mold once it's built.

With all of the pouring and flowing, the drip tray at the bottom should be full with a layer of silicone. If it's not at least ½″ (1 cm) thick, go ahead and mix up another small batch and fill the tray until it is.

With this last bit of silicone poured on, leave it alone so it can cure. Seriously. Stop touching it.

When the rubber has cured, use a pair of pliers to twist the pegs and pull them out of the silicone around the bottom edge (Figure 8-15). The cardboard drip wall can be removed, as well.

Building the Mother Mold

At this point, even though it's nice and thick, the rubber jacket mold would collapse or warp under its own weight if the prototype was removed. To prevent this, you'll need to build a support structure to hold it in the proper shape and keep everything straight. Functionally, this is the same sort of mother mold that you built for the two-part molds in the second part of Chapter 7.

Once again, fiberglass is ideal for this application.

Before building up the fiberglass, you'll need to do something to make sure you can take it apart afterward. This calls for a parting wall to establish the edges of the parts of the mother mold. To build the parting wall, pick a line that will neatly divide the piece in half. Make sure that the rigid shell built up on either side of that line will be able to be neatly removed without breaking or tearing any of the rubber jacket mold. Once you've determined the best way to split the mother mold in half, use soft oil-based clay to build up a wall, as shown in Figure 8-16.

FIGURE 8-16: The clay parting wall

Once the parting wall is built up and smoothed out, it's time to start building the fiberglass shell. A mother mold like this calls for at least three layers of lightweight fiberglass mat. Any less and there will be gaps in the strands of glass that will make weak spots. Any more and the mother mold will be unnecessarily heavy and use more material than it really needs.

Put on some disposable gloves. Long sleeves are a good idea too.

Start by tearing the fiberglass mat into patches about the size of your hand. Set a pile of these patches next to the work area.

Next, tear off some smaller patches and stuff them into the holes around the bottom edge where the pegs used to be (Figure 8-17).

PUT ON YOUR RESPIRATOR. This is the part that's going to get stinky. The stink is bad for you. Nobody will be impressed if you think you're just tough enough to inhale a bunch of poison, so knock that off.

FIGURE 8-17: Peg holes loosely stuffed with fiberglass

Once you've torn off enough fiberglass patches to cover the entire side of the mold three times, mix up about a pint (about half a liter) of fiberglass resin in accordance with the manufacturer's instructions.

🔫 Maker Note

The mixing instructions will call for a certain amount of catalyst that's ideal for nominal atmospheric temperatures in the workspace. On hot days, the resin will cure faster. On cold days the resin will cure slower. The ambient temperature can be compensated for by adding a couple of drops more or less of the catalyst to make the resin cure faster or slower respectively.

Once the resin is mixed, you have a limited time to use it (known as *pot life*) before it will begin to gel up and become impossible to apply with a brush. At this point it's time to quit screwing around, put down the cell phone, and focus on working quickly. It's a good idea to take a bathroom break before mixing the resin. That's not the kind of emergency anybody wants to handle with sticky, resin-coated fingers.

Using a disposable chip brush, coat the surface of the mold with the mixed fiberglass resin (Figure 8-18), making sure to coat all the way out to the edges of the clay parting wall. Along the way, fill the peg holes with resin, too.

FIGURE 8-18: The mold coated with resin

While the resin is still wet, start placing the fiberglass patches into it (Figure 8-19). Be careful to ensure that the strands lie flat against the surface and don't get folded up or jammed into one another. This will make it harder to work out the bubbles later.

Once you have a few patches of fiberglass mat laid into the resin, use the chip brush to add more resin on top of the mat. Use a brushing motion to force the resin into the mat and push any bubbles out of the resin.

Since the area against the parting wall will be the only part that has any real structural requirements, it's a good idea to start the fiberglass layup around the edges, as shown in Figure 8-20.

FIGURE 8-19: Fiberglass patches laid into the wet resin

FIGURE 8-20: Putting fiberglass on the edges of the mold first ensures that nobody will forget to get them covered.

Then fill in the area in the middle (Figure 8-21). Done properly, there should be no bubbles trapped in the fiberglass mat. If there are, keep working them out with the bristle brush.

Once the fiberglass on the first half has cured, it's time to take a look at the other side (Figure 8-22).

Remove the carefully crafted clay wall you put so much work into earlier (Figure 8-23). There may be a tiny bit of residual clay left behind. Scrape off as much as possible. Eventually you'll just want to settle for smoothing it out the best you can.

FIGURE 8-21: Fiberglass laid up across the rest of the mold

FIGURE 8-22: By the time you're looking at this angle, you're halfway done with the mother mold.

FIGURE 8-23: The second side of the mold with the clay parting wall removed

FIGURE 8-24: *Clay blob.* That's the technical term.

🔫 **Maker Note**

Use a small amount of clay to fill in the gap between the fiberglass and the silicone. This will keep the resin from the second half of the mother mold from seeping in between them and gluing the two halves together.

Add a couple of blobs of clay along the edge (Figure 8-24). These will give you a place to insert a screwdriver or other prying tool to separate the halves of the mother mold later.

Before laying up the fiberglass for the second half of the mother mold, something must be done to ensure that the two halves won't bond together. While there are a lot of great release agents available, this can be easily done with a thin layer of petroleum jelly. Just scoop a goober of it onto a fingertip and spread it evenly over the entire flange area, as shown in Figure 8-25.

Once the flange is evenly coated, it's time to lay up the fiberglass mother mold for the second side (Figure 8-26). The process is the same as with the first side. Remember to force some of the mat into the peg holes, glass the outside edges first, then cover the whole thing with at least three layers of mat.

FIGURE 8-25: Goobering the petroleum jelly onto the flange with a fingertip

FIGURE 8-26: Meet the second half; same as the first half.

When the second half of the mother mold has cured, drill a few ¼" (6 mm) holes around the outside edge, just far enough from the body of the mother mold to fit a bolt and a wingnut onto it (Figure 8-27).

FIGURE 8-27: Holes drilled around the edge of the flange

Now at last we come to the time of the unveiling! Turn the whole arrangement on its side and remove the cardboard that was attached to the bottom, as shown in Figure 8-28.

FIGURE 8-28: The cardboard on the bottom. It's gone!

Next, insert a screwdriver or other suitable prying instrument into the slot provided by the clay blob wedges along the edge of the flange (Figure 8-29).

With a bit of prying, the two halves of the mother mold should separate cleanly, as shown in Figure 8-30.

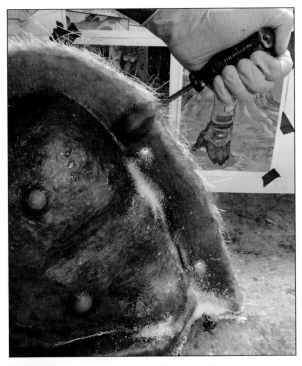

FIGURE 8-29: Screwdriver gently stabbed into the unsuspecting clay blob

FIGURE 8-30: The great divergence, when what was once one is torn asunder

With a bit of jiggling and wiggling, the mother mold should be able to slide off the rubber jacket without any damage (Figure 8-31).

If the silicone rubber is unable to stretch far enough to easily remove the prototype, use a sharp hobby knife to make the relief cut along the back of the helmet as shown in Figure 8-32.

FIGURE 8-31: The mother mold removed from the rubber jacket

FIGURE 8-32: The squiggly relief cut

🔫 Maker Note

The relief cut is not at all straight. This is to make sure that the two halves are properly aligned when it comes time to reassemble the mold for casting. If they're not put together right, the jacket mold will never fit into the mother mold correctly, liquid resin will leak out of the split in the mold, the cast part will be twisted and misshapen, there will be fire and brimstone coming down from the skies, rivers and seas boiling, 40 years of darkness, earthquakes, volcanoes, the dead rising from the grave, human sacrifice, dogs and cats living together, mass hysteria . . . Okay, maybe not quite all that, but it's worth getting it right.

Grab an edge of the silicone rubber jacket mold and peel it off the prototype (Figure 8-33). Be gentle. There's no good reason to stretch the rubber any more than absolutely necessary. The more stress it goes through, the sooner it will wear out.

FIGURE 8-33: Removing the prototype from the rubber jacket

Before you put the mold back together, it's a good idea to trim the sharp, hairy, or ugly bits off the mother mold. This step isn't entirely necessary, but your dainty soft fingertips will appreciate it later.

The ideal tool for trimming the edges is a rotary tool with a cut-off wheel (Figure 8-34).

FIGURE 8-34: The cut-off wheel

Wear safety glasses and a dust mask. This tool is designed to make short, straight cuts, so it takes a steady hand to avoid breaking it and having little pieces of cutting wheel explode all over the place. Wearing safety glasses ensures that you'll just have to pick the broken bits out of your hair. It also functions by grinding its way through the material being cut. This generates quite a bit of dust. The dust mask keeps your lungs relatively dust-free (Figure 8-35).

Once the edges of the mother mold have been trimmed, the rubber jacket can be placed back into one half; then the other half can be set in place and the whole arrangement can be bolted back together (Figure 8-36).

FIGURE 8-35: Imagine the gray funk this guy would be coughing up if he wasn't wearing a mask! (The hat and coveralls are actually bright orange under the dust.)

FIGURE 8-36: The reassembled mold; ready to roll

Now that the mold is made, it's time to crank out a casting.

Rotocasting

For this stage, you'll need to either lay down a drop cloth or relocate to a work space where there won't be any ill consequences when liquid nastiness drips out of the mold and turns into funky-colored plastic permanently bonded to the floor. The clothes you wear during this process will be ruined, as well. You've been warned.

The rotocasting process is pretty straightforward. To begin, mix a batch of resin in accordance with the manufacturer's instructions. For a snug-fitting helmet like this one, you'll need about 8 fluid ounces (approximately 250 mL) of resin to get a good, thorough coat. You'll want to mix this first batch by hand, instead of using a mechanical mixer, to avoid introducing any excess bubbles. Using a mechanical mixer will also cause the material to cure faster, which may or may not be a good thing.

Once the resin is thoroughly mixed, pour it slowly into the mold, taking care not to let it splash around (which would add more bubbles). Then, pick up the mold and do the rotocasting dance (Figure 8-37).

The object is to spread the liquid resin over the entire surface of the inside of the mold. Look inside while you're doing this. If the resin has started to pool in one particular spot, roll the mold over so that spot is now on top. If there's a high point where the resin hasn't flowed, turn the mold however you need to in order to get the whole thing coated.

Eventually, the resin will begin to thicken and gel as it cures. This will make it harder to get it to flow around the inside of the mold. At this stage you need to be very careful to avoid having any big blobs that will become excessively thick spots, or drips that will turn into sharp stabby bits pointed at the wearer's head. Keep the mold moving until the resin has stopped flowing. When it cures, it will change color, typically becoming somewhat lighter as it solidifies, as shown in Figure 8-38. At that point, it should look nice and smooth inside.

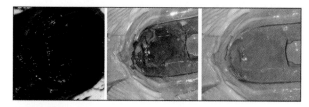

FIGURE 8-38: Going, going, gone. The casting resin changing color as it cures.

FIGURE 8-37: To the casual observer, you have now lost your darned fool mind.

While that first coat might be nice and smooth inside, it's probably still pretty thin. It's time to repeat the process to make the shell thicker.

🔫 Maker Note

When urethane resin cures, it may end up with an oily residue on the surface. While this is not at all unusual, it will prevent subsequent layers of resin from bonding together. So once the first layer has cured enough to stop flowing, it's time to mix up the next batch and start rolling it in the mold again. If you wait too long between coats, there's a chance they'll delaminate later, and the resulting casting will be very weak.

Somewhere between three and five coats should be enough to ensure that the entire shell has been built up with adequate strength for costume use.

🔫 Maker Tip

If you want a thicker shell without adding a lot of weight, add a few scoops of microballoons into the resin mixture. Microballoons are tiny, hollow glass spheres used to add volume to a composite without adding much weight. They can be purchased in many of the same places where urethane resin is sold.

After the last coat has had enough time to cure completely, the mother mold can be unbolted and removed, and then you can peel off the rubber jacket just like you did when you removed the prototype (Figure 8-39).

Now that you know the mold will work, the prototype can be discarded. You don't need it anymore. Instead, you can make all of the cast copies you could ever want (Figure 8-40).

FIGURE 8-39: The strong, lightweight, rotocast copy next to the fragile, heavy, pink prototype

FIGURE 8-40: Don't get carried away.

Trimming and Prepping the Cast Part for Painting

Now that you've got that cast part out of the mold, you may notice a few flaws that were created during the molding process. Most notably, the big flap of flashing (Figure 8-41).

FIGURE 8-41: Gah!

FIGURE 8-42: Whew!

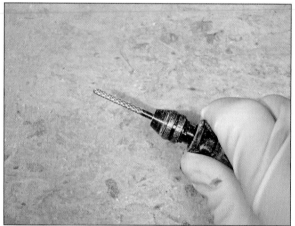

FIGURE 8-43: Designed for cutting ceramic tile, this bit does a great job of tearing through resin.

Fear not, in most cases, the flashing can be snapped off with your bare fingers (Figure 8-42).

Since you're already wearing clothes you don't care about, break out the rotary tool again and trim all the unnecessary resin off the casting. A lot of bits will work well for this job, but a tile-cutting bit (Figure 8-43) is a good place to start.

This particular bit is a little tough to control, though, so it should only be used to do the rough trimming of areas like the neck hole and the eye holes (Figure 8-44).

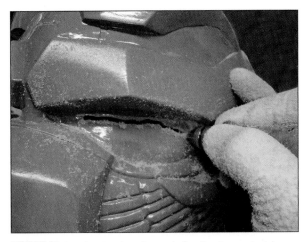

FIGURE 8-44: Rough trimming the eye holes, leaving material around the edges to be cleaned up later

Once the rough trimming is done, switch to a coarse sanding drum (Figure 8-45).

FIGURE 8-45: 80-grit sanding drum—because sometimes roughness is the best path to smoothness

The sanding drum will grind through a lot of resin pretty quickly, so be careful not to get carried away. By the time you're done, 90 percent of the trimming work should be behind you.

Eventually, no matter how much you dislike the notion, power tools will be a bad idea. At this point you can do the last bits of cleanup around the edges with progressively finer grits of sandpaper and some good old-fashioned elbow grease.

With the edges trimmed, give the rest of the casting a once-over to see if there are any little blisters on the surface (caused by bubbles in the silicone mold). These should be cut, sanded, or filed off.

Now that all of the excess has been trimmed off, sand the entire thing with a fine-grit sandpaper (at least 220-grit) to make it easier for paint to stick to the surface. Then wash the casting in warm, soapy water to remove any silicone or oily residue from the surface. Rinse it off with clean water, and allow it to dry. If you skip this step, surface contaminants will end up ruining the final paint job.

🔫 Maker Note

Dish soap is a great choice for this step. If it claims to be tough on grease, that's good. If it promises to destroy grease, then hunt down grease's entire family and murder them in their sleep in order to send a message to any other oily deposits that might get in its way, it may be a bit too harsh.

Now the cast part is ready for paint (Figure 8-46 on the following page).

See Part IV of this book for more about painting.

FIGURE 8-46: The fully painted helmet

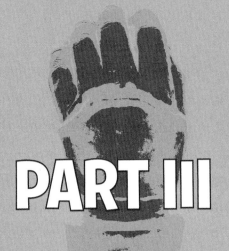

PART III

Vacuum Forming

9

VACFORMING BASICS
Make Factory-Looking Parts at Home

VACFORMING IS A GREAT way to make costume parts. Just a few decades ago, it called for specialized shops equipped with expensive industrial equipment and difficult-to-find materials. Nowadays, the process is so common that most people will throw away countless vacformed things every day without giving them a second thought. Vacformed parts are used for everything from refrigerator linings and product packaging to disposable coffee cup lids and hard-shell luggage (Figure 9-1). What's more, with just a few simple tools, you can vacform right in your own home.

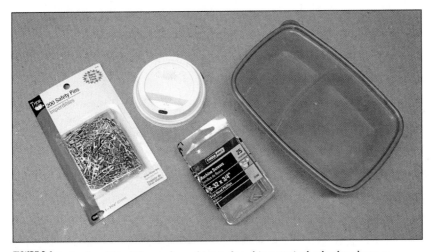

FIGURE 9-1: Just a handful of the vacformed parts found in a typical suburban home

155

How does it work? Here's the short version:

- Clamp a sheet of plastic into a frame.
- Heat the sheet until it is soft and flexible.
- Stretch the framed plastic sheet over a form.
- Suck the air out from under the plastic sheet.
- Once the plastic has cooled and become rigid again, remove it from the form.

The cooled plastic will retain the shape of the form.

Warning

You must take great care to avoid catastrophe throughout this process. Be awake and alert, and pay attention. If you get distracted during the vacforming process, you can ruin appliances, burn down your home, fill the air with poisonous fumes, and light yourself on fire.

Sound like fun? Let's get started . . .

Terminology

A few technical terms will come up during this chapter that you should become familiar with right off the bat:

VACFORMING   Also called *vacuforming*, or *vacuumforming*, if you're not into the whole brevity thing. This is the process of using heat and vacuum pressure to form sheets of heated plastic.

FORMING BUCKS   Also called *forms*, *bucks*, or occasionally *molds*, these are the shapes that will be used to form the sheets of plastic.

PLATEN   The surface that the forming bucks are placed on. It is usually a box that is perforated on top and attached to a vacuum source that will draw the air out from underneath. Rhymes with *flatten*.

CLAMPING FRAME   A pair of wooden or metal rectangles that the sheet of plastic gets clamped between.

VACFORMING MACHINE   A purpose-built contraption combining a heat source, frame, platen, and vacuum source.

Designing the Forming Bucks

When you're heating and bending plastic sheets by hand, you can make a few shapes without any kind of guide whatsoever. But when you start vacforming, you're going to need some shapes to start with. These are called *forming bucks*.

When designing the forming bucks, you should take a few things into consideration:

- Undercuts or overhanging areas of the forming buck will cause the plastic sheet to lock in place, making it impossible to get the plastic off in one piece.

- Tall or deep parts will require the plastic to stretch farther to cover the forming bucks. The farther it stretches, the thinner it will be.

- If the shape to be formed has a deep recessed detail that will trap air once the plastic seals against it, no amount of sucking will pull the plastic into that area. Consider drilling tiny holes to allow the air to flow out of that area through the bottom of the forming buck.

- There will be a tremendous amount of pressure on the bucks during the vacforming process. Make sure that they are sturdy enough to hold up.

Building the Forming Bucks

Forming bucks can be made of anything that will hold up to the heat and pressure of vacforming (Figure 9-2).

While any number of found items can be used as bucks, at some point you'll need to make something custom. How you make them is really just a question of what you feel comfortable working with.

You can design them in a CAD program and carve them out with a CNC machine. No CNC machine in your garage? Build them using Pepakura and just be sure to reinforce them for extra strength. If you can work out the shape in your head, and then pick up a hammer and chisel and sculpt them in marble, go for it. It's really just a matter of what's available and your personal skill set.

If you don't have a personal skill set, don't fret. The easiest way to make a forming buck is to pick up a block of balsa wood at your local hobby shop, or green florist foam from the local craft store, and then cut, sand, or file it down until it's the shape you need (Figure 9-3). This will be more than adequate if you only need one or two copies.

If you need more than a handful of copies, take the first formed copy, remove the forming buck, and fill the formed sheet with plaster. When it cures, you can pop it out and use this new sturdier plaster form instead (Figure 9-4).

That's the easy way. Another option for more complex shapes is to sculpt them in clay. If you can get the rough shape in clay, it's a simple thing to reproduce it in plaster and use the copy for a forming buck. First, you'll need to gather a few things.

What you'll need:

- Oil-based clay such as modeling clay or plasticene
- Water-based clay
- Sculpting tools

FIGURE 9-2: A few examples of things that can be used for simple forming bucks

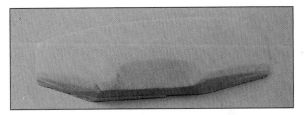

FIGURE 9-3: A block of green florist's foam whittled down to the shape of Hunter's lower chest armor

FIGURE 9-4: Pulling a plaster copy of the forming buck from a formed sheet

- Gypsum, such as Hydro-Stone, Hydrocal, or UltraCal. These are available in better art supply stores or shops that specialize in ceramics. Plaster of Paris will work in a pinch, but it won't be as sturdy or long-lasting. From now on, we'll just call it *plaster* for the sake of brevity.

- Mixing buckets

- Burlap

- Chip brushes

- Acrylic clearcoat in a spray can

- A mold release such as petroleum jelly

- Someplace you can make a mess

Step 1

Sculpt the part out of oil-based clay. Figure 9-5 shows the rough shape of another piece for the Hunter costume. There are countless ways to learn how to sculpt, but they all boil down to one simple method:

- Grab some clay.

- Take away the bits that are sitting where they shouldn't be.

- Add bits where they're missing.

- Repeat until you have the shape you want.

  The only way to get any good at it is to practice.

Step 2

Spray on a coat of clear acrylic. This will form a barrier between the plaster and the clay and help to prevent the plaster from adhering to the clay.

Step 3

Determine if your part needs to be broken down into more than one piece. Figure 9-6 shows a good example of a part that needs to be split up.

If there are overhangs and undercuts, it's time to set up a *parting wall* that will become the separating line between the two halves. The parting

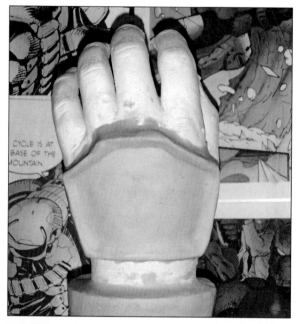

FIGURE 9-5: Hunter's hand plate sculpted in oil-based clay

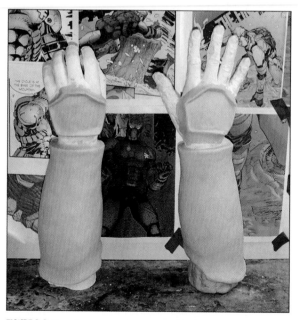

FIGURE 9-6: There's no easy way to get this forearm armor formed in one piece without the plastic sheet locking onto the forming buck.

wall can be made out of cardboard, wood, plastic, or sheet metal, but if the pieces being molded are relatively small, you can get away with just building a barrier out of water-based clay. An example is shown in Figure 9-7.

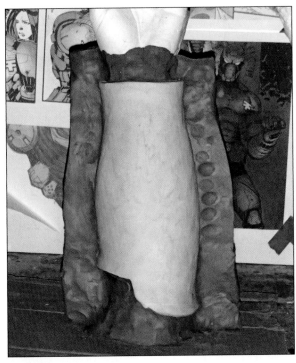

FIGURE 9-7: A parting wall made of water-based clay

🔫 Maker Note

The parting wall will need to have a watertight seal against the sculpted clay part. Cut it as close as possible to fit onto the oil-based clay sculpture, then use water-based clay to "caulk" the edge. Then, coat the parting wall with petroleum jelly or some other release agent to keep the plaster from sticking to it.

Step 4

Once the parting wall is in place, mix up a small batch of plaster. Use a chip brush to gently paint it onto the surface of the clay sculpture. This will be the *print coat* that will pick up all of the surface details (Figure 9-8).

FIGURE 9-8: The plaster print coat

Step 5

After applying the print coat and before it has completely cured, mix up a thicker batch of plaster and layer it on top to make the mold thicker, as shown in Figure 9-9. Take care to avoid trapping any air bubbles.

FIGURE 9-9: Thickening the mold with more plaster

Mixing Plaster

When you buy gypsum products, they'll usually come with instructions to tell you the exact best way to add water. Oftentimes they'll actually tell you to measure out the powder and water by weight, but let's face it: you likely won't be measuring this stuff on a scale.

If you're more of a visual person, here's a quick-and-dirty how-to.

Starting with a matched pair of clean containers, add a few inches of clean, room-temperature water to one and an equal amount of plaster to the other.

Pour in about half of the plaster powder. It should make a neat little island in the middle (Figure 9-10).

Stir the island into the water using a mixing stick, drill mixer, or better yet, your bare hands. Be sure to eliminate any lumps or dry chunks along the way. This means repeatedly scraping the sides and the bottom of the container to blend in all of the material.

For building up thickness, you'll want the mixture to be thicker to avoid having it run off and drool all over the place. If the mixture is too thin, add a bit more powder. If it's too thick, add a bit more water. For print coats, the mixture should be just a bit thicker than milk. Try to get it about the consistency of melting ice cream.

FIGURE 9-10: Plaster Island—land of the Gypsum people

Step 6

For especially large pieces, it's often a good idea to reinforce the plaster for extra strength. You can use plaster-soaked strips of burlap (Figure 9-11) or plaster bandages from the local pharmacy to strengthen the mold.

Step 7

Allow the piece to cure fully before you mess with it any more.

FIGURE 9-11: Plaster-soaked strips of burlap laid into the still-wet plaster will make for a stronger mold.

Step 8

If the part required a parting wall, now it's time to remove it as gently as possible. Make sure to avoid bumping or scratching the clay sculpture.

Step 9

Add a couple of small clay tabs to the area where the parting wall used to be on the plaster, as shown in Figure 9-12. Then apply a light coat of petroleum jelly to prevent the second side of the plaster mold from sticking to the first side.

Step 10

Repeat steps 4–7 for the second side.

Step 11

After the plaster has cured, pry the two halves apart by inserting a pair of screwdrivers or chisels where the clay tabs used to be, as shown in Figure 9-13.

Step 12

Close off the open ends of the two molds, as needed. You're going to have to fill them with liquid plaster, so cap off the ends by gluing on pieces of cardboard or plastic sheeting (Figure 9-14).

FIGURE 9-13: Using a pair of screwdrivers to gently pry apart the two halves of the plaster mold

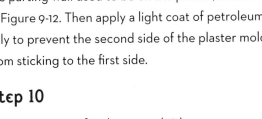

FIGURE 9-12: Clay tabs added to the parting wall will provide starting points when it comes time to pry the two pieces apart.

FIGURE 9-14: The open end of the molds closed up with corrugated cardboard and hot glue

Step 13

Coat the insides of the molds with petroleum jelly. Cover everything, but don't make it so thick that it covers all of the details.

Step 14

Fill the mold with plaster, as shown in Figure 9-15.

FIGURE 9-15: The molds filled with plaster

Step 15

Pop the cast part out of the mold. Since you're only making this one copy, don't worry if you have to break the mold to get the cast piece out (Figure 9-16).

Step 16

Sand the plaster piece as needed to get it nice and smooth. You can also carve in any small grooves or details you want to show in the vacformed part (Figure 9-17).

FIGURE 9-16: The mold has done its duty and shall be forever remembered.

FIGURE 9-17: The finished forming buck after sanding it smooth

Now that you have some shapes to form, it's time to get your sheets together . . .

Choosing Plastics for Vacforming

Styrene, ABS, and PETG are the most common types of plastic sheets that are used by costume hobbyists. In some cases acrylic sheet is also used.

STYRENE Also known as *polystyrene* or *HIPS* (*high-impact polystyrene*), it is the same stuff that most model kits are made of. It is available in a wide variety of colors and usually has a matte surface finish. It is easy to heat, cut, sand, and paint.

ABS Short for *acrylonitrile butadiene styrene*. Years ago, some mad scientist, determined to rule the world of plastics, mixed a couple of extra materials into styrene and made a stronger, more flexible plastic. ABS sheets come in a variety of colors and are usually glossy on one or both sides. Compared to styrene, it's a little tougher to get paint to stick to ABS, but the benefits in strength and flexibility more than make up for this.

PETG Less commonly known as *polyethylene tere-phthalate glycol-modified*, this is basically the same plastic that is used to make soda and water bottles. PETG is transparent and relatively inexpensive. It can also be dyed to create custom-tinted parts. Much like ABS, it can be a bit tricky to heat it prop-erly for vacforming, but once you get used to its quirks it's a great option.

ACRYLIC While a few rare nerds in the plastics industry will try to impress you by calling it *poly-methyl methacrylate*, normal human beings are more likely to refer to this stuff by brand names such as Plexiglas, Acrylite, Lucite, or Perspex. It is available in a wide variety of colors (clear or opaque), and the surface comes with a nice, glossy finish.

While it is more rigid than PETG, acrylic can be a huge pain to work with. The surface is porous and microscopic pinholes will collect moisture. This isn't usually a concern when it's being used as a window. But when the sheet is heated up, these bits of water will boil and expand, causing countless tiny bubbles to form on the surface of the flexible heated plastic. To avoid this, the material needs to be dried out by baking it at a low temperature for a while before you attempt to get it hot enough to vacform.

ABS tends to heat up unevenly and end up with hot spots that will ruin the formed piece. PETG and acrylic require careful handling and storage that will cause a lot of headaches if you're not familiar with the quirks.

For first-time vacformers, there's nothing bet-ter than styrene. Starting out with styrene will save you time and money during the learning stages.

Most towns with more than a few thousand people living in them will have some sort of local plastic supplier. If you can't find one locally, a quick web search for "sheet plastic" will take you to all sorts of websites for companies that will offer all of these materials in a variety of thicknesses cut to size and shipped right to your front door.

In the United States, suppliers will usually list the sheet thicknesses by *gauge*, measured in thou-sandths of an inch. For example, 0.060" (called *sixty gauge*, *sixty mil*, or *oh-sixty*) will be a bit less than 1/16" thick. For durable, wearable armor parts, you'll want to use plastic that's at least 0.080" and preferably more like 0.100".

If you're blessed with the metric system, there are no magic words to describe the thickness of the material. Shop for something at least 2 mm thick and it'll all work out.

The Simplest Possible Vacforming Setup

If you have access to a makerspace, tech shop, or a friend with a professional vacforming machine like the one shown in Figure 9-18, you'll save yourself a lot of trouble. But if there's nothing available in your area, don't despair. You can get by without it.

FIGURE 9-18: This old workhorse of an industrial vacuum former, now cranking out parts at 32Ten Studios, built an entire army of Stormtroopers back in the 1980s.

🔫 Maker Note

A typical household vacuum cleaner probably won't generate enough suction to do the job. For a project like this, use the highest horsepower shop vacuum you can get your hands on.

If you access to an oven and a powerful vacuum cleaner, you're halfway there. Add a homemade platen and clamping frame and you'll be ready to get cooking.

Build the Clamping Frame

The frame can be made out of whatever you're comfortable working with. Just bear in mind that it'll need to withstand high heat when it's placed in the oven to cook the plastic sheets. It can be wood, but it's better to use a hardwood such as oak, poplar, or walnut. If you make it out of something cheap such as pine, it'll be more likely to burn.

You can also make the frame out of aluminum angle stock, available from your local hardware store.

Remember that you will want to measure and build the frames to fit the oven that you will be using. Make it as big as it can possibly be, while still fitting into the rack ledges inside the oven with the door closed.

A wood frame held together with metal brackets is more than adequate for vacforming. A longer-lasting option is a metal frame made of welded or riveted aluminum or steel, but as long as you're paying attention (and provided you're not working with Satan's own oven), a frame like this will never burn to the point of being useless.

Start by making the largest wooden rectangle that will fit into your oven (Figure 9-19).

Screw an angle bracket onto each of the corners (Figure 9-20).

Now make another copy of the exact same rectangle (Figure 9-21).

FIGURE 9-19: Wooden parts for one side of the vacforming frame

FIGURE 9-20: A bracket. Screwed.

FIGURE 9-21: A matched pair of frames

Once you've made a matched pair of rectangles, you'll need to work out a way to hold them together to trap the plastic sheet in between them. One easy way to do this is by drilling holes along the edge and fitting them with bolts and wingnuts. You may be tempted to add a hinge on one end, but this will be a bad idea when you have to clamp sheets of varying thickness from one project to the next.

🔫 Maker Tip

If the holes aren't evenly spaced along the edge of both frames, mark the edges so you know which way they're supposed to fit together. That way you won't spend all day getting frustrated when they don't fit together because you're assembling them backward.

If you'd like to make a smaller frame to handle smaller sheet sizes, it doesn't mean you'll need a smaller oven. It just means you'll need to work out a method to suspend the smaller frame, as shown in Figure 9-22.

FIGURE 9-22: A small baking tray can be used to suspend the frame and keep the drooping, heated plastic sheet from sticking to the racks in the oven.

The Forming Platen

In essence, the platen is just going to be a box. It needs a fitting on the side or bottom where a powerful vacuum cleaner can be connected to it. Then the top needs a bunch of holes drilled in it so that it can suck in air. Think of an air hockey table that's stuck in reverse.

The exact dimensions of the forming platen will be determined by the inside dimensions of the clamping frame. Properly sized, the platen should fit inside the frame with a gap of about 1/8" (3 mm) or so all the way around.

Once the size is determined, cut out a couple of plywood rectangles that will fit inside the frame. Set one aside. This will become the bottom of the box.

By making the forming platen fit inside the frame, the heated plastic will end up forming an air seal all the way around the edge when it is pulled down over the platen. It's a good idea to round off the top edges of the platen with a router or even sand it by hand in order to make a nice, soft edge. This will give the plastic more area to seal against and make it easier to pull the formed sheet off the platen once it has cooled and become hard again.

For the top, start by marking a grid on the surface, as shown in Figure 9-23.

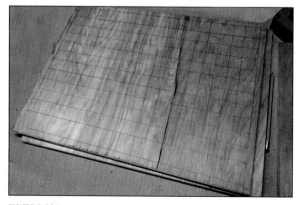

FIGURE 9-23: The grid

Now drill a hole at every intersection (Figure 9-24).

FIGURE 9-24: The grid drilled

🔫 Maker Note

The top of the platen doesn't necessarily need to have a lot of holes, but the more there are, the less thought will have to go into making sure that air will flow under the forming buck. More holes will make it easier to evenly distribute vacuum suction across the entire surface of the platen. Smaller holes will make it less likely that something will get sucked inside of the box.

With the holes drilled into the top of the platen, it's time to add the sides of the box, as shown in Figure 9-25. These can be glued or screwed in place, but bear in mind that the box seams will need to be airtight, so it can be a good idea to do both. The height of the forming platen isn't all that important. A couple of inches (a few centimeters) should be more than enough.

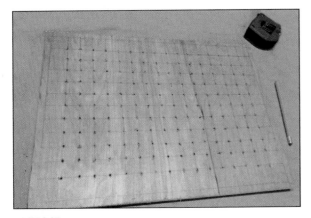

FIGURE 9-25: Box sides added

🔫 Maker Note

Someday in the future, it may become a good idea to upgrade to a stronger vacuum source than your shop vacuum. If this is a possibility at all, it is a good idea to add two or more supports in the middle of the platen to help it hold itself up against the additional pressure. You may have noticed a pair of these supports in Figure 9-24.

Now it's time to add a hose fitting to plug the vacuum hose into. A hose adaptor (Figure 9-26) will serve nicely.

FIGURE 9-26: Hose adaptor

Drill a hole big enough for the small side of the hose adaptor to fit into, then glue and screw it into place, as shown in Figure 9-27.

With that done, it's time to close up the bottom with that other plywood rectangle. Then, just to be sure, it's a good idea to seal the edges again. This may be one of the only instances where duct tape is used to guard against air leaks.

Finally, add a descriptive label so everyone will know what this contraption is for (Figure 9-28).

FIGURE 9-27: Hose adaptor installed with glue and screws to make it airtight

FIGURE 9-28: With the edges taped and a name added to the box, the vacforming apparatus is complete.

Now that the clamping frame and forming platen have been made, it's time to move the whole operation into the kitchen (Figure 9-29).

FIGURE 9-29: The forming platen plugged into the shop vacuum and positioned conveniently near the oven

With that whole setup assembled, it's TIME TO COOK! But first, be sure to don all of the necessary protective equipment, as shown in Figure 9-30.

Actually, all you really need is a pair of gloves or oven mitts for handling the hot stuff coming out of the oven (Figure 9-31).

Now you need to fit a sheet of plastic into the frame. If you can't get your supplier to cut the sheet to the exact size you need, it's simply a matter of scoring a line with a sharp knife and a straight edge, then snapping it along that line (Figure 9-32).

FIGURE 9-30: This is overkill.

FIGURE 9-31: More than adequate protective equipment (not pictured: optional apron)

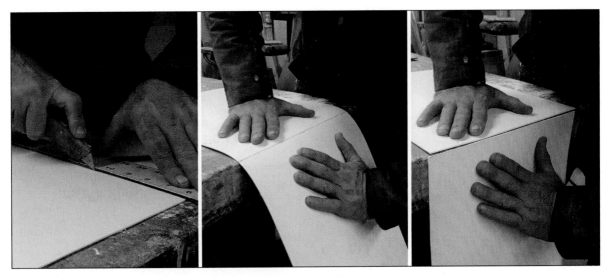

FIGURE 9-32: Scoring, bending, and snapping the sheet to get the just-right size

In this case, the plastic sheet is sized to fit in between the bolts on the frame. Once the plastic is in place on the frame, bolt it shut (Figure 9-33).

FIGURE 9-33: Clamping frame with plastic sandwiched inside

Turn the oven up to 350°F (around 180°C) on the bottom burner. Make sure you're not using the top burner (usually turned on in "preheat" and "broil" settings) because this will be more likely to burn the sheet of plastic and cause the whole thing to heat up too quickly and unevenly.

While waiting for the oven to heat up, place the forming bucks on the platen.

🔫 Maker Note

The placement of the forming bucks requires some careful consideration. Too close together and the plastic won't be able to stretch into the space between them without rupturing or being drawn into thin ridges, known as "webbing" (Figure 9-34). Too far apart and there will be a lot of plastic wasted in between the parts. There is often no "correct" way to lay parts out. Experience is the ultimate teacher.

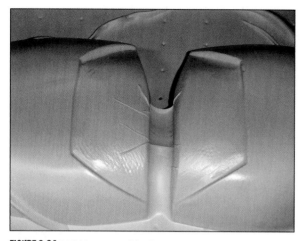

FIGURE 9-34: Webbing caused by having two forming bucks too close together

If you haven't done it already, **put on a pair of gloves!** From here on out, the process is going to involve touching hot stuff. Don't kid yourself; you're nowhere near callous enough to handle the hot plastic and clamping frame with your bare hands. If this doesn't make sense, the friendly staff at the local burn ward will be happy to explain it to you.

If you have a heat gun handy, now's a good time to plug it in and place it within reach.

Once everything is handy and the oven is warmed up, slide the clamping frame into the rack slots in the side of the oven (Figure 9-35). Pick a slot somewhere in the top third of the oven so it's not too close to the burners on the bottom.

FIGURE 9-35: The clamping frame inserted into the oven

> ### 🔫 Maker Note
>
> Be sure that the frame isn't going to slip out of the slots in the oven. The last thing you want at this stage is for the whole arrangement to flop down onto the hot burner and catch fire. It'll smell bad.

Stay there and watch while the plastic heats up. At first it won't seem like anything is happening. Then, slowly, the surface will begin to warp. It'll form waves, ripples, or rolling, undulating hills and valleys across its surface (Figure 9-36). This is happening because certain parts are absorbing a bit more heat than others and the warmer, softer parts are expanding, while the cooler, harder parts are not. It's normal.

Next the plastic will flatten out. At this point it is all evenly heated and somewhat flexible. Keep it in the oven a bit longer. Things are *about* to get interesting.

Finally, as the plastic continues to heat up it will begin to sag under its own weight. This is where timing will become very important. The

FIGURE 9-36: Wavy, warm plastic, as foretold in the Prophecy.

object is to melt the sheet to some extent. Allow it to sag too much, and you run the risk of the whole thing melting its way onto the burner and starting a fire. Allow it to sag too little and it won't be stretchable enough to pull down over the forming bucks. What's more, as it absorbs more heat, it will actually start to sag faster the farther down it gets. Again, experience will be the best teacher here. Plan on having a few extra sheets of plastic to account for the pieces that'll be discarded during the learning process.

Now things are going to really suck.

Once the plastic is sagging down in the middle by a couple of inches (a few centimeters), turn on the shop vacuum and pull the frame out of the oven. Quickly transfer the plastic to the forming platen. Do this as quickly as possible so the plastic doesn't have enough time to cool and become stiff again.

Push the clamping frame straight down onto the platen (Figure 9-37). When the plastic meets the edges of the platen, it will form a seal, and the air will be sucked out. This will cause the hot plastic to shrink and wrap itself around whatever is sitting on the platen.

FIGURE 9-37: The forming

If there are areas that don't get pulled into details on the forming buck, you can use the heat gun to warm up an area locally. Try to spread the

🔫 Maker Note

When forcing the frame down over the platen, push straight down. If it's pressed down crooked, the plastic will be stretched more on one side than the other. The side that stretches farther will end up being thinner. In this context, another word for thinner is *weaker*. Weaker is bad.

heat around so that the plastic will stretch as uniformly as possible; otherwise a small hot spot will become a soft area that will stretch too thin and possibly pop, ruining the whole piece.

If the vacuum doesn't suck enough to pull the plastic into the details or inside corners, you can also use your hands or some tools to force the plastic into place. Again, wear gloves or oven mitts.

Keep the vacuum turned on until the plastic has cooled and become rigid again. Depending on how thick the sheet is, this could take anywhere from 20 seconds to about a minute. If life moves too fast and there isn't time to wait, use a fan or a compressed air nozzle to blow cool air across the platen and speed things up.

Once the sheet has become rigid again, shut down the vacuum. The formed sheet may make a few popping or cracking noises as the pressure is released. Don't panic. This is normal.

Now it's time to pop the forming bucks out of the formed sheet of plastic (Figure 9-38).

🔫 Maker Note

In many cases, this just means picking up the clamping frame and shaking lightly to get the forms to fall out. For more stubborn parts, it may mean pushing down on the center of the part while lifting up on the edges of the plastic. For the most stubborn pieces, you may need to use a utility knife or a pair of shears to make "relief cuts" to get the bucks out.

Once the plastic is removed from the frame, you're ready to do it all again. With a bit of practice and careful planning, you'll have a suit of armor in no time at all (Figure 9-39).

FIGURE 9-38: The formed sheet—success!

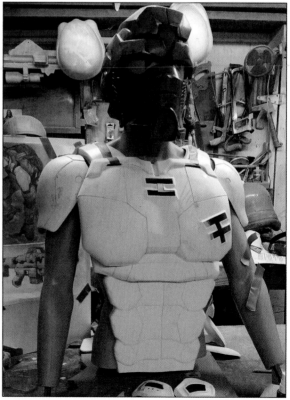

FIGURE 9-39: Hunter's armor coming together

Maker Tip

Instead of trying to force the knife blade through the plastic, lightly score the surface where you want the cut to be, then go back over it repeatedly with the knife until you cut through. This will give you a lot more control and be a lot safer than trying to hack all the way through with one pass of the blade.

Forming Multiple Layers to Add Fine Details

You'll notice that the parts pictured have some very fine seam lines. While it is possible to get these small details by carving them into the forming bucks, it will mean that the plastic will have to be pretty thin in order to fit into these small details. If you need seam lines like these, the better option is to vacform a sheet over the forms as we've already demonstrated, then trim around the bucks, leave the formed part on the buck, and form another sheet of plastic right over it. Remove both layers from the forming buck, cut your seams and small details in the outer layer, and glue the pieces back together with the visible gaps in the top layer making your seams. Figure 9-40 shows the pieces after each step.

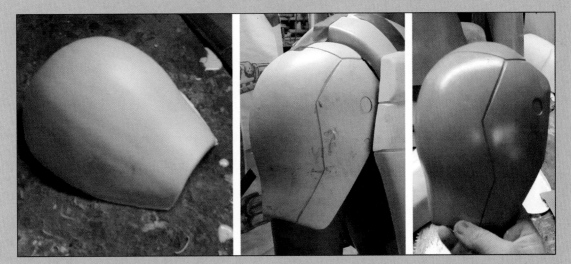

FIGURE 9-40: Hunter's shoulder formed as a single layer of plastic (left), two layers with details cut into the outer layer (middle), and glued together with paint and weathering (right)

Once you have all of your parts formed, trimmed, and glued together, it's just a matter of painting them, which is covered in Part IV of this book, and everything will be ready to strap together, as shown in Figure 9-41.

FIGURE 9-41: The complete set of parts for Hunter's armor, all painted up and ready to go

BUILD YOUR OWN VACUUM FORMING MACHINE

Step Up Your Game by Stepping Out of the Kitchen

THE LAST CHAPTER EXPLAINED the minimalist approach to vacforming by using your kitchen oven and a vacuum cleaner. While that's all well and good, eventually you'll need something else. Maybe you'll need a dedicated system that you can set up in your shed or garage that's ready to go in a moment's notice. Maybe the other people who eat foods prepared in your oven will complain about the subtle taste of melted plastics that permeates everything. Maybe you've burned down your house and want to spend the insurance settlement on better equipment for your hobbies.

Whatever the reason, just as a true Jedi master must build his own lightsaber, sooner or later you'll level up to a whole new mastery of prop-making abilities and build your own vacformer. You'll need a basic understanding of electricity (or a friend who has it). You'll also need some woodworking tools (or a friend who has them).*

---

*Or an enemy who has them and spends a lot of time away from home.

Warning

Everything about this chapter is bad for you. When vacforming in the kitchen, you were already constantly on the edge of burning down your house and filling your lungs with all manner of poisonous fumes. Now you'll have the added risk of electrocuting yourself with jolts of potentially heart-stopping voltage while building a custom oven or cannibalizing an electric heater for parts. You'll be taking apart a perfectly safe, UL-listed appliance and turning it into some kind of Frankenstein's contraption. Study up on your electrical safety procedures, but at the very least:

● Unplug electrical devices before you start touching wires and the like.

● As always, if you don't know what you're doing, get help from someone who does.

Begin by Building the Clamping Frame

In Chapter 9, the dimensions of the clamping frame were determined by what would fit in the oven. Now that the oven is no longer a concern, it's time to figure out what size clamping frame is right for you. Since the clamping frame will drive the design of the rest of the machine, it's a good idea to start here.

There are two main things to consider:

● Standard sheet stock sizes

● Available heat sources

● How much space is available to store the machine

Three things. Oops.

Sizing for Standard Sheet Stock

Here in the United States, sheet goods like this are typically sold in 48″ × 96″ pieces. This being the case, it's wise to build your clamping frame to hold a piece that can be evenly cut from a sheet this size. For example, a 24″-square frame means that you can cut that full sheet into eight usable pieces. Meanwhile, a 20″-square frame still gets you eight usable pieces and a bunch of oddball scraps on the ends. Unless you want to wallpaper your workshop with rectangular bits of plastic, these scraps will be wasted.

While bigger can often be better, bear in mind that there is also such a thing as too big. Imagine if you somehow managed to build a machine that could vacform an entire 48″ × 96″ sheet in one piece. While that might be a great way to impress your prop-making friends at parties, pretty soon you'll find yourself in a spot where all you need is one little widget and you have to use up an entire sheet to form it. When it comes to making armor and prop pieces, 24″ square is often large enough.

Sometimes smaller is better, as well. It's easier to handle the unwieldy clamping frame single-handedly, and you don't end up wasting a bunch of material if you only need to form something small.

Sizing for Available Heat Sources

The other main consideration when choosing the size of the clamping frame is going to be the size of the heat source you'll use. In this case, we'll be using a repurposed electric heater. The heat will radiate away from the heater, and the farther the plastic is from the heat source, the less heat will actually reach it.

As you can imagine, using an electric frying pan to heat your sheet is fine for a square piece of plastic that's about the same size as the pan. You may even be able to make the clamping frame a bit bigger or smaller. But if the sheet is much bigger than the frying pan, you'll end up with a hot melted spot in the middle of the sheet, and the outside edges will never get warm enough to form.

Let's make a clamping frame to hold sheets that are 12″ × 16″. Why 16″? Because 48″ divided by 16 gives us three equally sized pieces. This means that a full sheet can be cut into 24 equally sized pieces with no waste at all.

A frame this small can be easily made with some simple hand tools and then riveted together. Figure 10-1 shows the tools and materials you'll need.

FIGURE 10-1: A drill with a ⅛″ bit, a ruler, a permanent marker, a pop rivet gun, 90° corner brackets, pop rivets, washers, ½″ aluminum C-channel stock, and a hacksaw

🔫 Maker Note

A miter box will help greatly. Otherwise, a carpenter's square and a steady hand will have to suffice.

Now that you've gathered all of the tools, start by cutting out two sets of aluminum channel pieces (Figure 10-2). This means two pairs of 12″-long pieces and two pairs of 16″-long pieces. The cuts on the ends need to be miter-cut with a 45° angle in order for the corners to fit neatly together.

Next, fit one of the corners together and use an angle bracket to ensure that it is square, as shown in Figure 10-3.

FIGURE 10-2: Two complete sets of frame parts

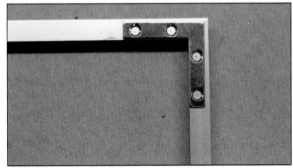

FIGURE 10-3: A corner fitted with an angle bracket

Mark the screw holes with a pencil or permanent marker, as shown in Figure 10-4.

Then drill a hole for each mark (Figure 10-5).

Once all of the holes are drilled, use pop rivets to hold the pieces together (Figure 10-6).

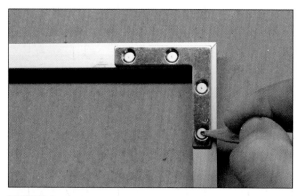

FIGURE 10-4: Marking the locations to drill

FIGURE 10-5: Drilling the frame

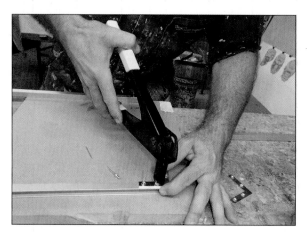

FIGURE 10-6: Riveting

Repeat this operation for all four corners on both sets of parts and the clamping frame is done (Figure 10-7).

Now it's probably plainly obvious to even the most casual observer that this clamping frame is really just a pair of metal rectangles. In fact, while it's barely a frame at this point, snarky commenters will point out that it doesn't actually have any way of doing any kind of clamping at all. They can be instantly silenced with the installation of a series of binder clips, as shown in Figure 10-8.

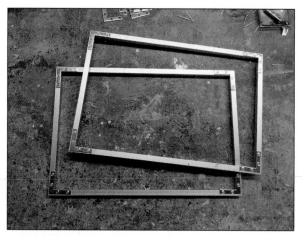

FIGURE 10-7: Boom! Clamping frame

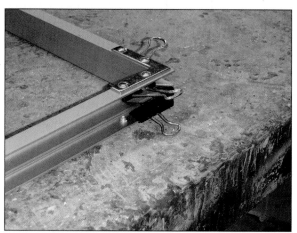

FIGURE 10-8: Binder clips make a pair of rectangles into an actual clamping frame.

Once there are clips attached all the way around the outside edge, the frame is ready to be fitted to a cut sheet of plastic (Figure 10-9).

Now you just need some plastic, a heat source, a forming platen, and a vacuum source.

FIGURE 10-9: The frame, ready and waiting

Heat Sources

There are a lot of good options when it comes time to create the heater to cook the plastic. All of them are electric. It has always been this way.

> *Throughout the course of human history, electricity has found no better use than the heating of plastic sheets for the purpose of vacuum-formation into suits of armor for cosplaye.*
>
> —Benjamin Franklin, 1779 (probably not really)

Sure there are a lot of other things that can get hot. Looking around the garden shed or junkyard, a crafty forming-machine builder might be tempted to use an old propane heater or possibly a charcoal grill. The more scientifically minded might have a steam radiator in the bedroom that seems ripe for hacking. More environmentally conscious costumers may be tempted to harness the heating power of the sun by using some sort of parabolic mirror to heat their plastic. The more nature-friendly among you may think of using a campfire to cook up the sheets for forming.

All of these are terrible ideas. Remember, when heating up sheets to form, it's vital to be able to reliably control and evenly distribute the heat being applied to the sheet. Anything that involves actual flame will inevitably involve hot spots and have a greater risk of actually catching the sheet on fire.

So while scavenging for parts, look for things like hot plates, toaster ovens, or electric frying pans. If you are a bit more comfortable working with electrical tools, space heaters can be repurposed and reconfigured for this application. If you're even more educated in the mysteries of electron flow, you can make your very own heating elements by just running a current through a coiled length of nichrome wire. There are also readily available quartz and ceramic heating elements that are designed and sold as plug-and-play components, allowing buyers to arrange numerous pieces together to make ovens in any shape or size they need.

Continuing our example build, we'll be cannibalizing the household electric space heater shown in Figure 10-10.

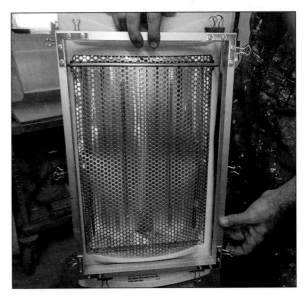

FIGURE 10-10: An otherwise ordinary electric space heater that's about the right size for the clamping frame

Before committing to a particular heat source, it's important to verify that it actually generates enough heat to make the plastic formable (Figure 10-11).

Common Configurations

Much of the reason for building a homemade vac-forming machine is to have a completely integrated system that can be turned on and used without having to go through the trouble of cleaning out an oven, setting up the vacuum, and otherwise rounding up tools and materials in order to get the job started. Instead, all of the parts will be attached to each other and self-contained. This can be done in a variety of ways, as shown in Figure 10-12.

When it comes time to attach the heat source to the rest of the machine, there are generally two ways to go about it: either side-by-side with the

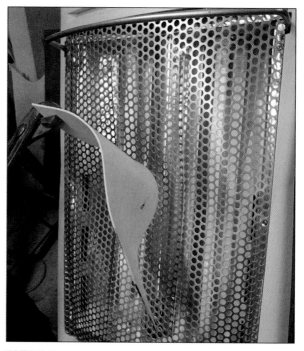

FIGURE 10-11: Warped, sagging, melted plastic means it's good enough.

FIGURE 10-12: Just a few examples of homemade vacforming machines

heat source mounted adjacent to the platen, or an over-and-under configuration with the heat source mounted above the platen.

Side-by-Side

With this configuration (Figure 10-13), the clamping frame ends up being placed above the oven during heating. After the sheet is hot enough, it is moved over to the platen for forming. Oftentimes, the frame is actually mounted in place with a hinge arrangement to keep things neatly aligned, but this means that the sheet of plastic will have to be flipped over when moving it onto the forming platen.

This configuration has one great advantage: heat rises. This makes the heating more efficient and makes the cooking stage of the vacforming a bit faster.

This configuration also has one great disadvantage: melting plastic falls. This means that if the operator gets distracted and the sheet is allowed to sag too far, it's almost guaranteed to start a fire. It'll smell bad when the plastic starts burning. There are some other reasons you probably don't want a

fire in your workspace, but suffice it to say that this configuration is less than ideal.

Over-and-Under

With this configuration, the biggest challenge is finding a way to hold the clamping frame into place below the oven. This can mean adding notches, clamps, hooks, or magnets so that the clamping frame is held to the heat despite the ever-present clawing hands of gravity.

Other than that, there are no real disadvantages to this setup. There is a little heat loss due to convection (the reason heat rises) but it's insignificant compared to the increased safety afforded by having the heated plastic fall away from the heat source in the event the machine is left unattended (Figure 10-14).

For the demonstration build, the plastic body is removed from the space heater (Figure 10-15).

Taking care to keep all of the electrical connections intact, the sheet metal reflector, heating elements, and wiring are then mounted into a

FIGURE 10-13: A typical, homemade side-by-side vacforming machine made by Sean Bradley of Sean Bradley Studios

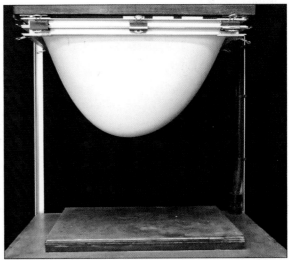

FIGURE 10-14: This overcooked sheet is absolutely not about to burst into flame.

properly sized hole cut into a piece of plywood (Figure 10-16).

Since the heater is a little smaller than the clamping frame, the frame will need to be held a short distance away from the heating elements. This will allow the heat to radiate further outward and evenly heat the entire sheet. This is accomplished by building a box around the heating elements, as shown in Figure 10-17.

At this point, each corner of the stand-off is fitted with a 90° angle bracket, spaced to match the brackets on the clamping frame (Figure 10-18).

Finally, sides are added to the whole assembly so that the wires and controls can all be contained (Figure 10-19).

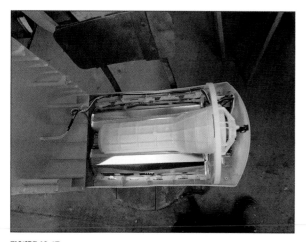

FIGURE 10-15: The back shell of the space heater removed

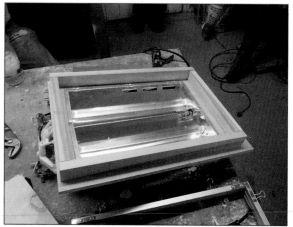

FIGURE 10-17: The stand-off box

FIGURE 10-16: The hot stuff screwed onto a piece of wood

FIGURE 10-18: Brackets mounted to the corners of the stand-off box

FIGURE 10-19: The completed heater enclosure. The cutout in the front is where the controls will be installed later.

Building the Platen

The forming platen will be built pretty much the same way as the platen that was built for forming in the kitchen. The main difference is that it will be sized differently in order to match the inside dimensions of the clamping frame.

Start by cutting out a plywood rectangle that's large enough to fit inside the clamping frame, with no more than a 1/8″ (4 mm) gap all the way around.

Once the plywood has been cut, it needs holes in it for the air to pass through. At this point, it's time for another design decision. How many holes?

Technically, the forming machine will work if there's just one large hole in the middle somewhere. The downside of this setup is that care must be taken to ensure that air can flow under the forming bucks and into the one suction hole. The upside is that it'll still work if you decide to set up a smaller clamping frame and you don't want to have to make another, smaller platen just to fit it.

On the other hand, by drilling a bunch of smaller holes, the platen will be more forgiving

when it comes to the layout of all of the parts. Smaller holes will also keep larger chunks of debris from getting sucked into the vacuum source, which could potentially cause damage.

Finally, the forming platen needs to be made of relatively sturdy material. Atmospheric pressure at sea level is 14.7 psi (101.325 kilopascals). This means that if you've built a platen that's 10″ × 10″ (23 cm × 23 cm) then sucking all of the air out from inside the platen will theoretically require it to hold up under 1,470 lbs (about 2/3 of a metric ton) of pressure. So don't cheap out and think a thin piece of perforated particle board will do the trick. Instead, get a nice, sturdy piece of plywood.

For the demonstration machine, the platen was cut from a piece of 1/2″ (about 10 mm) shop plywood. Then it was marked with a grid. The line spacing was about 1″ (22.5 cm) vertically and 1.25″ horizontally (Figure 10-20), mostly for aesthetic reasons.

Once the grid was laid out, 1/8″ (3 mm) holes were drilled at every intersection, as shown in Figure 10-21.

FIGURE 10-20: The grid marked on the surface of the platen

With all of the holes drilled, a piece of ¾" MDF was cut to fit the bottom of the platen, forming a border framing the whole thing. Then the MDF was glued in place, as shown in Figure 10-22.

Finally, another piece of plywood was cut to the same dimensions as the piece the heater was mounted to, and then glued to the bottom of the platen (Figure 10-23).

There was also a hole cut into the bottom piece, as shown in Figure 10-24. This was cut with a hole saw.

Finally, add sides to the bottom of the platen to build a base that everything will be mounted onto. The sides need to be tall enough to fit whatever vacuum source will be used to drive the machine.

Vacuum Sources

There are a lot of options when it comes time to source the sucker.

Keep an eye out at yard sales for a working Shop-Vac that can be disassembled for parts. A

FIGURE 10-21: Drilling holes

FIGURE 10-23: The bottom of the platen glued in place

FIGURE 10-22: The border glued in place on the bottom of the platen

FIGURE 10-24: A hole

disembodied vacuum motor (Figure 10-25) is a great source for suction.

Another great option is a vacuum pump (Figure 10-26). These can be found at bigger stores that specialize in selling tools and industrial equipment. The main advantage of a vacuum pump is that it will generate a much higher vacuum than anything else you can buy. The disadvantage is that it'll be costly and usually will involve lower flow volume. This means that it'll pull the parts harder, but it'll take longer to remove all of the air from under the sheet.

If you're a fan of manual labor, a bicycle tire pump can be modified by attaching a hose from the platen to the intake side of the pump. It'll take a bit more physical work for each pull, but it's an option.

Not fond of pumping by hand? A small air compressor can also be used by simply fitting the suction hose to the intake side of the compressor.

Finally, sucking on the hose with your mouth can typically generate as much as 15"Hg worth of vacuum. It'll be hot and it'll taste bad, but it's more

than adequate to pull detailed parts, so it can be an option for a small machine. It will be a good idea to install a valve on the hose somewhere so you won't have to stand there sucking on the hose until the plastic cools.

Once you've obtained the vacuum source, it's just a matter of building it into the base under the platen. Figure 10-27 shows a vacuum motor mounted snugly over the hole in the bottom of the platen. An additional hole was cut in the side of the box to allow air to flow out of the back of the machine during use.

FIGURE 10-26: A small vacuum pump designed to service automotive air conditioning systems

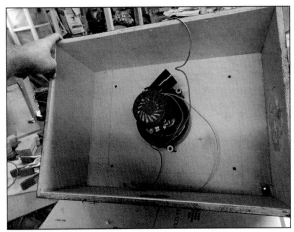

FIGURE 10-27: The vacuum motor installed in the bottom of the platen

FIGURE 10-25: The driving force behind the vacuum cleaner

Once that's done, the base was fitted with a set of struts made of aluminum angle stock, and the heater box was attached to the top, as shown in Figure 10-28.

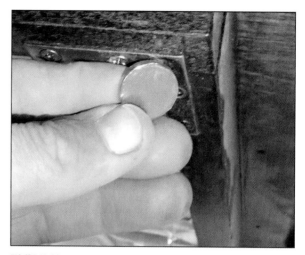

FIGURE 10-29: Magnets hold the clamping frame up to the heater.

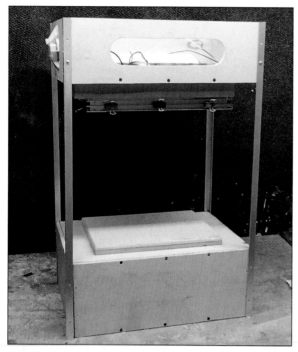

FIGURE 10-28: The assembled forming machine as a completely self-contained unit

FIGURE 10-30: The slightly dressier finish for the vacformer

Maker Note

A set of four rare earth magnets is used to hold the steel angle brackets on the clamping frame onto the steel angle brackets on the heater enclosure, as shown in Figure 10-29.

At this point, it's probably a good idea to add whatever decorative elements the machine will need, as shown in Figure 10-30. This is purely optional, but if you're going to build your own personal machine, you might as well personalize it.

Finally, it's time to do all of the wiring. It's a good idea to lead the power for the vacuum motor and the power for the heater to a common On-Off-On toggle switch (Figure 10-31). This way you can be sure you've turned the heater off before turning the vacuum on. There's no sense in having the thing crank out heat while you've got your hands in there.

FIGURE 10-31: The toggle switch to select between turning on the heater and turning on the vacuum

In this case, all of the controls were mounted into the cutout in the front of the heater enclosure (Figure 10-32).

Maker Note

Electric space heaters will usually have a shut-off switch of some sort designed to make sure the heater turns itself off if it falls over. When building a machine like this, the shut-off switch must be removed, bypassed, or otherwise disabled. They're wildly different from one heater to the next, so if you don't know what you're looking at, make friends with an electrician so they can help you out.

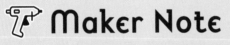

FIGURE 10-32: A piece of bottle green acrylic makes for a handsome mounting surface.

Finally, use tape, hooks, or wiring staples to hold all of the wires in place away from the hot parts, as shown in Figure 10-33.

Maker Note

For even more detailed instructions on how to build customized vacforming machines, check out http://www.build-stuff.com/ where you can buy plans and instruction manuals to build either of the machines pictured in Figure 10-34.

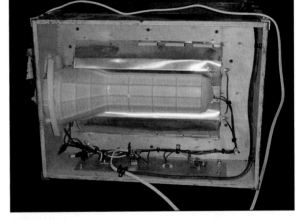

FIGURE 10-33: Tying down the wires makes sure they won't slip into the wrong place and start some kind of meltdown.

FIGURE 10-34: A Hobby Vac machine built by Liz Hydra of Hydra-works and a customized Proto-former built by Harrison Krix of Volpin Props, both based on plans from build-stuff.com

Putting It to Work

So there it sits—a self-contained machine standing by to take on all of the vacforming chores, whatever they may be.

To use it, start with a sheet of plastic that's cut to the same size as the outside edges of the clamping frame. If your local supplier isn't able to provide you with the right-sized sheet, don't fret. Most of the plastics you'll be forming for costume use can be easily cut with a utility knife and a straight edge, as shown in Figure 10-35.

Sandwich the sheet inside the clamping frame and clamp it shut with the binder clips (Figure 10-36).

Now set the clamping frame in place under the oven, using the rare earth magnets to hold it in place (Figure 10-37).

Turn on the heater and wait. Note that the heater has an adjustable thermostat. After a few pulls, you'll start to get a good idea of the best heat settings for various types of plastic. It's a good idea to mark these settings directly onto the control dial (Figure 10-38).

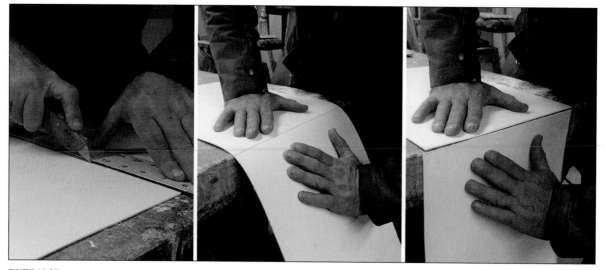

FIGURE 10-35: To cut the sheet, simply score it, then bend the scored line over a hard edge to snap it.

FIGURE 10-36: The plastic sheet in the clamping frame

FIGURE 10-37: With the frame in place, it's ready to cook.

FIGURE 10-38: Have some fun with the labeling.

Once the plastic has sagged enough to make
the pull, flip the toggle switch from heat to vacuum.
Then pull the clamping frame down from the oven
and press it over the platen. Figure 10-39 shows a
successful pull on this machine.

Now you can do all of your vacforming in the
garage or backyard without having to tie up the
kitchen oven or make a lot of noise in the house.

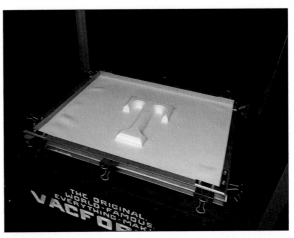

FIGURE 10-39: SUCCESS!

PART IV

Painting and Weathering

11

BASECOATS AND MASKING

Building a Believable Finish Starts with a Solid Foundation

LET'S SAY YOU'VE BEEN toiling away for a couple of weeks' worth of late nights. Your workspace is covered with bits of paper, dust, and scraps of plastic; your fingers are worn; and sweat drips off your furrowed brow. But somehow at the end of all this you've emerged victorious. You have made a thing (Figure 11-1).

FIGURE 11-1: Things!

The problem is, at this point, the thing probably looks a lot like the paper, dust, scraps of plastic, and floor mats that are scattered all over the rest of your workspace. As luck would have it, the wise and ancient peoples of the world have foreseen this very problem. Indeed, you have fulfilled the exact prophecy that caused the great minds of antiquity to invent paint! Their forethought makes it possible to turn these mere things into phenomenal-looking props, such as those in Figure 11-2.

But what is paint? Well . . .

FIGURE 11-2: The same things, but with paint

More Than You Ever Wanted to Know About Paint

To put it in simplest terms, paint is a liquid that can be spread in a thin layer over an object before turning into some kind of solid film. Typically paint is made up of three basic elements: binder, solvent, and pigment.

The binder is the liquid everything else is mixed into. It usually determines finished properties such as sheen, flexibility, and durability. The solvent serves to dissolve the various solid elements and adjust the viscosity of the paint while wet. Typically the solvent will evaporate as the paint dries and does not become part of the resulting finish. Pigments are the ground-up solid bits of color.

There can also be other fillers blended in to make the paint tougher, add texture or other special surface properties, or just to reduce the cost of the paint (like when manufacturers add sawdust to your breakfast cereal). But ever since ancient man started mixing colored dirt with tree sap and water (the pigment, binder, and solvent, respectively) and

spreading it on cave walls, most paints have pretty much consisted of those three basic elements.

Of course, these days there's not a lot of reason to mix up your own paint. If you live in any kind of civilized society, ready-made paint can be found waiting in the nearest hardware, hobby, or art supply store.

Types of Paint

ACRYLIC This is a fast-drying paint containing pigment suspended in a polymer emulsion. Acrylic paints are water-soluble but become water-resistant when dry. Acrylic paint can be diluted with water, or modified with acrylic gels, media, or pastes to change its dried properties. Available in gloss or matte finishes, acrylics are wildly versatile and retain some flexibility when dry. This makes them ideal for surfaces that will be subject to bending.

ENAMEL These are oil-based paints, usually with a significant amount of gloss in them. These are usually paints with a hard surface and of higher quality, including floor coatings of a high-gloss finish and most commonly available spray paints. Enamel paints air dry to a hard finish, and are used for coating surfaces that are outdoors or otherwise subject to hard wear or variations in temperature and humidity. The hard surface tends to crack or flake when applied to flexible surfaces.

LACQUER This is a clear or colored finish that dries by solvent evaporation to produce a hard, durable finish. This finish can be of any sheen level from ultramatte to high gloss, and it can be further polished as required. It typically dries better on a hard, smooth surface.

URETHANE This is commonly used for automotive finish applications, urethane paints are very durable. Properly maintained, a coat of urethane paint will outlast most other paints. With the right equipment and a bit of practice, they are easy to apply and provide excellent coverage. Urethane paints are activated by mixing in a hardener. This means that once the paint is mixed it must be used or it will be wasted, but this also means that it dries quickly. Typically urethane paint is applied in two stages with the color coat or coats being covered by a protective clearcoat, but there are also "single-stage" urethanes where the color coat is designed to hold up to exposure to the elements. The biggest drawbacks are the need to use a spray gun or airbrush to apply the paint and the fact that they are highly toxic and require careful precautions to avoid exposure to nasty airborne compounds that can be absorbed through the lungs and skin.

EPOXY Not long after two-part epoxy resins were invented for adhesive or composite uses, someone came up with the great idea of making colored versions for paint. Two-part epoxy coatings were originally developed for heavy-duty service on metal parts such as appliances and industrial machinery, so they provide a very durable, hard surface finish. They also give off toxic fumes, smell horrible while curing, and tend to be pretty expensive.

Paint Compatibility

Most paints rely on solvents to make them liquid enough to brush or spray onto a surface. Since many of these same solvents are often used by themselves to dissolve or remove paints, it's important to be sure that each successive layer isn't going to destroy the layers below.

As a general rule of thumb, the faster a paint dries, the more volatile its solvents are. It's important to allow each coat to dry thoroughly before applying another coat over it. Even then, generally speaking, you don't want to spray any kind of lacquer over an enamel paint. This is because the solvents in lacquers are more volatile than enamel solvents. This will cause the lacquer solvents to dissolve the enamel paint layers underneath and completely ruin the finish.

As with all things, there are exceptions to every rule. It is possible (although very tricky) to apply lacquer on top of enamel successfully. It's risky, but it can be done in very light, thin coats. By building up thickness gradually and allowing each coat to dry before applying the next one, it'll be possible to avoid putting enough solvent onto the enamel surface to cause the paint to lift or dissolve.

Enamels tend to have less volatile solvents and can be applied safely over lacquer in most cases. Water-soluble paints such as craft acrylics can be applied over almost anything.

Still, paint formulations have become so varied and complex in recent years that it's difficult to be certain whether one type of paint will be compatible with anything else. You may find that you have problems applying successive coats of enamel in different colors from one manufacturer even after adequate drying time between coats. At the same time, other paints will end up working fine no matter how many mistakes you make.

The best option in any case is to keep a few pieces of scrap material on hand to do paint tests. Each time you're going to add another coat to your project, start by adding a coat to the test piece to make sure the paints aren't going to eat each other and ruin all of your hard work.

Application Processes

When it comes to applying paint to an object, there are basically three options: solid, liquid, or gas. Seriously.

The solid version involves dusting an object with a heavy layer of fine-ground, dried paint and then baking it onto the surface. This is usually called *powder coating*, and will not really be useful for prop-making since it requires industrial equipment and high temperatures.

The liquid version involves dipping a brush, roller, rag, sponge, or greasy fingertips into a container of paint and then spreading it onto the surface. This is a great option that requires few specialized tools or skills. The main drawback is that it takes a great deal of care and a steady hand in order to ensure an even, smooth surface.

The gas version involves mixing tiny droplets of the paint into the air (often referred to as an *aerosol*) and then propelling that mixture toward the surface to be coated. When the paint hits the surface, it sticks. This is spray painting. This method

usually results in a smooth, even surface. It used to require specialized equipment, but now you can buy prepackaged spray cans with their own built-in spray nozzles and propellants that get disposed of after they're empty. Given the low cost, ease of use, and ready availability of spray paint cans, it's hard to see any drawbacks.

Basecoats: Building a Believable Finish Starts with a Solid Foundation

Before you begin, know where you're going to end up. Is the finished piece going to be coated with a thick, heavily textured paint? Then save a bit of time in the sanding stage if the paint is going to cover a few fine scratches. Does it need to have a glossy finish? If so, you're going to need to have it perfectly smooth before you begin. Start smooth in order to finish smooth. If you spray on a coat of primer, it'll help you notice minor flaws and scratches that'll need filling or sanding.

Surface Preparation

For a good-quality finish, 95 percent of the work is done before you apply even a single drop of paint. The parts need to be smooth, clean, and dry. If they're made of wood or reinforced Pepakura models, you'll need to be sure to seal the surface with primer in order to get an even, smooth finish.

Once you've got your parts sanded smooth, make sure the surface is free of dust and dirt and any kind of residue that might cause the paint to react adversely.

Prepping Cast Resin Parts for Paint

Resin cast parts need a bit of extra care in order to ensure you end up with a good paint job. After you remove any flashing, sprues, vents, or warts left behind from the molding and casting process, it's a good idea to lightly sand the entire surface with a very fine sandpaper (220-grit should be fine enough) in order to remove the smooth gloss from the surface and give it a bit of "tooth" for the primer to grab onto.

Look over the surface and apply spot putty or filler to any small pinholes or bubbles. Once that's dry, sand it smooth.

Finally, before applying any primer or paint, wash the parts in warm, soapy water to remove any remaining oily residue left on the surface of the parts by the silicone mold.

Preparing the Space for Painting

Once you have your parts ready for paint, you'll also need to get your work area ready for paint. Here are a few things to keep in mind:

Overspray Protection

When you are spraying paint, it will end up in a lot of places you don't want it to go. If you're only painting small parts, cover the area where you're working with newspaper. Tape it down in order to keep it from blowing around in the breeze as you spray. If you're spraying larger parts or working in a large area, don't waste time taping together newspapers. Instead, use a plastic or cloth drop cloth. Be sure to use one that's large enough for you to move the spray can past the object. If you're working outdoors, keep the wind in mind, and make sure you're not going to have a mist of paint droplets blowing onto your dog, your neighbor's kids, or someone's brand-new car (Figure 11-3).

Ergonomics

Place the piece you plan to paint so that it'll be easy to reach. This may mean using sawhorses or a workbench, or hanging it from overhead. If at all possible, don't set the work pieces on the floor. For one thing, you'll have to work hunched over

FIGURE 11-3: Possibly not the best place to start spraying paint

or kneeling. Worse than that, it'll make it harder to spray paint from all angles and get good coverage.

Mounting Small Objects for Portability

Whenever possible, place the object you're painting on a piece of cardboard or a scrap piece of plywood. For really small parts, mount them on a nail or piece of wire, like the pieces shown in Figure 11-4. This way you'll be able to move the object as you paint without touching it. If you have a lot of small parts, use a small turntable or lazy Susan to make it easy to spray them, then rotate the base to shoot paint from different angles.

Spray Can Handles

Most hardware stores carry aftermarket handles like the one shown in Figure 11-5. These are designed to snap onto the top of a spray cans to make them more comfortable to operate. The pistol grip shape and built-in trigger will help you get better results and prevent the fatigue you'd get from holding the can in your bare hand and poking at the spray tip with your finger.

Breathing Healthy

None of the solvents coming out of a spray can are good for you. WEAR A RESPIRATOR. A dust mask is not a respirator. It's also a good idea to work in an area that has plenty of ventilation.

Dress for Success

There are two schools of thought on this subject. Since you're going to probably end up with a few stray drops of paint on your clothes, it's a good idea to wear something you don't care about to prevent any sudden angry laundry emergencies. On the other hand, if you wear clothes that you care about, you're more likely to be careful while painting. This will mean making less of a mess in general,

FIGURE 11-5: A typical spray can handle makes spraying easier on your hands. The laser sight was added later by the author. For accuracy.

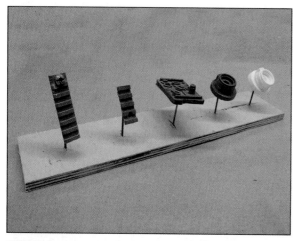

FIGURE 11-4: Small parts mounted on nails driven through a scrap of plywood

resulting in a better finish and easier cleanup. In the end, this book is not the boss of you. Wear what you want.

Primer

When it comes time to apply the primer, the first thing you should do is look at the can and READ THE DIRECTIONS.

That said, most primers will be pretty forgiving. Moreover, since this won't be the final finish, if you make a mistake and end up having to lightly sand off a drip or add another coat to cover a thin spot, it isn't the end of the world.

Primer is commonly available in four colors: black, white, gray, and red. The choice of color used to have to do with what kind of material it was

Brush Painting

The rest of this chapter will focus mainly on spray painting, because of its ease of use and ready availability. But if you're a traditionalist or you cringe at the thought of throwing away the empty cans when they're used up, here are a few quick tips for better painting with a brush:

- Before use, moisten the brush. Use water when painting with water-based paint, or turpentine with oil-based paint. Remove excess liquid before painting.

- Load up the brush by dipping the bristles no more than halfway into the paint.

- Tap the brush gently against the side of the can to remove excess paint. Don't wipe the brush hard against the lip.

- Don't press too hard. The bristles should flex only slightly as you brush. Paint should flow from the brush as you pull it along the surface.

- If you are painting a large area with a brush, use long, parallel strokes to cover it. Then brush across those strokes to even the paint out. Finish with another round of long, light strokes in the original direction.

- If you're painting a small part or picking out a tiny detail, try to place the heel of your palm against something that's fixed in place so that you're only using your fingers to move the brush, as shown in Figure 11-6.

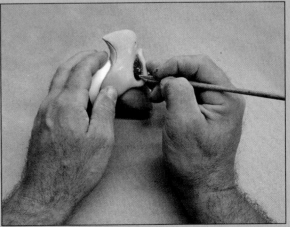

FIGURE 11-6: Resting your wrist or the heel of your palm against a stationary object helps to prevent the trembling caused by muscle fatigue so you can paint nice, straight lines.

going to be sprayed onto, but with modern paint it really doesn't matter. Now it's just a matter of picking the color that's closest to the color of the topcoat that will be applied over it. That way, if the topcoat is a bit thin or it gets scratched later on, the visible primer underneath won't be as glaringly obvious.

Applying Paint

Once the primer has thoroughly dried, it's time to add the paint.

Once you've picked out the paint you're going to use, the first thing to do is READ THE DIRECTIONS. If the manufacturer tells you to shake the can for one minute after the mixing ball starts rattling around inside, then that's a solid 60 seconds of your life that will be given over to shaking a can. If the directions state that you should hold the can upright and spray 10–12" inches away from the object being painted, don't start thinking you've got a better idea of how to get the job done. Chances are, the people who made the paint have been using it longer than you and already made all of the mistakes you're about to make. Learn from them.

Once you've read the directions, do a pattern test (Figure 11-7). Since different brands of paint will spray different-shaped plumes of paint, spray a test shot onto a piece of plywood to see the shape of the spray pattern that the can produces. This will allow you to adjust your technique and hold the can closer or further from the target, depending on the spray pattern.

Apply the paint in a sweeping motion; don't point and shoot. It's a spray can, not a camera. To get an even coat of paint, move the can horizontally and vertically past the object as you spray, as

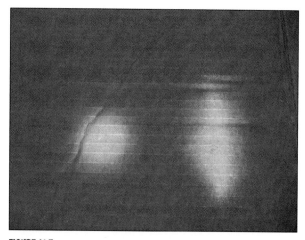

FIGURE 11-7: Spots of paint sprayed from two different cans showing their different spray patterns

shown in Figure 11-8. For example, if you're moving left to right, you begin spraying to the left of the object, onto the object, and keep spraying until paint is spraying off to the right of the object.

Oftentimes it's a good idea to spray a light coat all over the object you're painting. Then, before it has a chance to fully dry, go back and apply another light coat to fully wet the object with paint. Done right, this "wet-coating" method produces a smooth, flawless surface. Wet coating doesn't work with all paints, though, so try your method on a test piece.

FIGURE 11-8: Start spraying before the can is aiming at the object and continue after sweeping past.

Take care to be patient and do it right. Flaws can be fixed, but it's always better to just avoid them altogether. That said, a lot of things can go wrong with a coat of paint.

Common Paint Problems and Their Solutions

SAGS OR RUNS (FIGURE 11-9)   These occur when paint has been applied too heavily without having sufficient time to dry.

SOLUTION: This is where you get to practice patience. Stop trying to get the paint job done with only one coat! If you're doing it right, a good paint job will need at least two, and sometimes three, separate coats. If you get too eager and spot a run or sag, reposition the piece so that the run is sitting flat on top, then wait for it to dry. With a little luck, the run should level out and may completely disappear. If it's impossible to reposition the piece or you failed to notice the run until it dried, there's still hope. You just need to sand it smooth again and repaint it. This time with thin coats! Seriously.

BLUSHING (FIGURE 11-10)   Blushing is the result of moisture condensing on a cool surface while the paint is still wet. The result is a white haze, or dull, matte surface on otherwise glossy finishes.

SOLUTION: If you notice blush appearing as the surface dries, stop painting. It'll do you no good. If you can move the project to an area with less humidity or cooler temperatures, that may solve the problem and you can continue painting. If you've already coated the whole thing before the blush becomes apparent, you can always try polishing the surface after it has completely dried. Use a light buffing compound to reduce blushing and bring back the glossy shine. If that doesn't work, put on another coat of paint on a less humid day.

FIGURE 11-10: Blushing

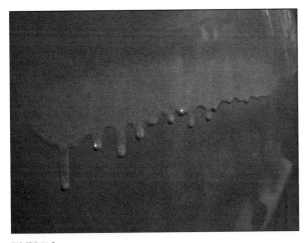

FIGURE 11-9: Too much paint!

BLISTERING (FIGURE 11-11) This is the formation of bubbles or bumps on the painted surface. There are two main causes for blisters: heat and moisture. Painting in direct sunlight or on a surface that is too warm can cause heat blisters. In essence, the surface of the paint dries too rapidly, forming a film that traps solvents underneath. The solvents eventually vaporize and create blisters or bubbles under the topcoat. Since darker colors absorb heat more than lighter colors, they'll be more prone to having this problem. Blistering can also be caused by any entrapped moisture in the object being painted. When it's cold, everything seems fine, but when it's warmed up, the moisture vaporizes and causes blisters under the paint.

SOLUTION: First, you need to figure out what's causing the blisters. If you pop one open and the bare plastic or wood is showing, it's probably caused by moisture. If you see another layer of paint under the blister, it's probably caused by heat. If moisture is the culprit, you need to remove all of the paint, reprime, and repaint. If it was heat, chances are the primer won't be affected and you can just sand the area smooth and repaint without priming. Just be sure to do it when the sun is not shining directly on the surface.

GRAINY OR BUMPY SURFACE FINISH (FIGURE 11-12) Often referred to as *orange peel*, this can be caused by a wide variety of factors. On a hot day, the paint can dry too fast, preventing the flying droplets from adequately blending together and smoothing out. If overspray lands on a previously painted area, it can leave small bumps behind. It could also mean the last coat you applied was too thin. Spraying while holding the can or airbrush too far from the surface allows the aerosolized paint to coalesce into bigger droplets that rain onto the surface, instead of a fine mist that will coat evenly. Improper storage of your paint may cause droplets of pigment to clump during application. This is typically caused by the paint freezing, or the can being too cold during application. Finally, it may simply be a function of the nozzle being partially clogged.

SOLUTION: If you notice this effect happening while you are painting, stop immediately. Allow the orange peel area to dry completely

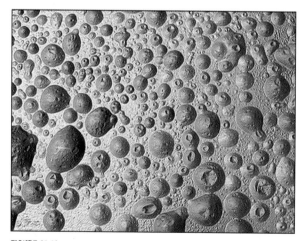

FIGURE 11-11: Blisters in the dry paint

FIGURE 11-12: This surface might be okay for citrus fruits, but not for a glossy coat of paint

before sanding it off with a fine-grit sandpaper. If it's warmer than about 70 degrees Fahrenheit (about 20 degrees Canadian), move to a cooler area. If you're using a spray can, be sure to shake the can thoroughly in order to adequately mix the solvents with the paint. Check to make sure that the spray nozzle is free of dried paint, dust, or other obstructions. If you're using an airbrush or spray gun, you may need to thin down the paint.

ALLIGATORING (FIGURE 11-13)   That is where the surface appears scaly or cracked like the skin of an alligator, hence the name. Alligatoring is often caused by an incompatibility between a topcoat and a basecoat. Oftentimes, it can be due to painting over a glossy finish that doesn't provide enough texture (occasionally called *tooth*) for the new paint to adhere to. Putting a hard finish coat over a soft primer coat can also be a cause. You can also see it if the undercoat isn't completely dry before recoating or if you apply too many coats of paint.

> **SOLUTION:**   Remove old paint completely by scraping and sanding, then repaint.

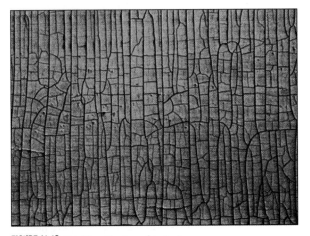

FIGURE 11-13: Crinkly alligator-skin texture

FISH EYES (FIGURE 11-14)   These are rings of thicker paint around an area where the paint is thinner or nonexistent. They are usually caused by contaminants, such as a spot of oil or silicone residue on the surface being painted. Frequently, you'll see this effect on a topcoat even after a primer coat has adhered perfectly to the surface.

> **SOLUTION:**   Sand the area around the fish eye until it is smooth; wash it thoroughly with warm, soapy water; swab the area with alcohol; allow it to dry; and repaint it.

FIGURE 11-14: These weird little craters are usually caused by surface contamination.

CRINKLING OR WRINKLING (FIGURE 11-15)   These can occur when the surface of a thick coat of paint dries before the rest of the paint underneath dries. The underlying paint shrinks as it dries, causing the surface to shrivel and pucker into a crinkled, wrinkly mess.

> **SOLUTION:**   This can be prevented by only laying down thin coats and giving the paint adequate time to dry in between coats. By the time the problem is actually apparent, you'll

need to remove the paint in the trouble area and repaint it.

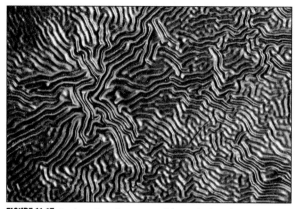

FIGURE 11-15: Puckered surface caused by painting the coats on too thick

BRUSH MARKS (FIGURE 11-16) These are visible streaks that remain after the paint has dried. This is usually due to using poor-quality brushes. Other causes include an overly porous surface that absorbs too much of the paint, excessive brushing, not applying enough paint, using the wrong thinner, or not allowing enough drying time between coats. You may also see brush marks when using a paint that dries too rapidly.

FIGURE 11-16: Brush marks

SOLUTION: Use higher-quality brushes (you cheapskate!). Make sure they have the correct type of bristles for the paints that you're using. Sand the trouble area before repainting; otherwise, the brush marks will still be visible through the new coat. Remember to paint multiple thin coats instead of one thick coat, and allow enough time for the paint to dry between coats.

HAIR, DUST, LINT, OR BUGS (FIGURE 11-17) Because sometimes things go wrong.

FIGURE 11-17: His friends called him Kamikaze Ken, and it was only a matter of time. So it goes.

SOLUTION: Stop painting. Recite the appropriate incantations and offer up this tiny sacrifice to invoke the benevolence of the Project Gods. Let the paint dry. Pick out the bits of dead bug (or fuzz or whatever) with a pair of tweezers or a sharp hobby knife, then sand the area smooth. Use long sanding strokes and keep as much of the paper flat against the painted surface as possible. You want to keep from grinding through the primer or digging out a "dish" in the finish. Wipe down the area with alcohol and begin the painting process again.

Masking and Stenciling

Chances are, your prop or costume pieces will need to have more than one color on them. After you've sprayed on your basecoat, it's time to add secondary colors.

Masking

Unless you are the master of all things spray can, chances are you're going to want to do something in order to keep the next coat of paint off certain areas. Before you start, though, make sure to allow enough time for the basecoat to dry completely. Once the paint is nice and hard, you can start adding masking tape.

Choose the Right Masking Tape

Masking tape technology has advanced by leaps and bounds in the last few decades. Make a quick trip to the local hardware store or local paint supplier and you'll be amazed by all of the masking tape options (Figure 11-18).

While there may be times when a cheaper product will do the job as well as the expensive product, cheap masking tape is almost never a good deal. You need tape that will seal tightly to the surface and come off easily without taking the paint off with it. The cheaper, tan-colored masking tape is usually aggressively tacky and will often leave behind some residual adhesive or lift the underlying paint when removed. For general masking, the blue painter's masking tape is usually adequate. There's also a gentler version of paper masking tape (usually colored purple) that's less sticky, which is gentler on more delicate surfaces. Both of these can be left in place for a week or two without doing any damage to the surface underneath. For finer details and sharper edges along curved surfaces, stretchable plastic masking tape is also available (usually from automotive paint supply shops).

FIGURE 11-18: A whole new world of masking tape options

Paint the Tape for Sharper Edges

If you're really worried about paint seeping under the edge of the masking tape, you can use your basecoat color to seal it, as shown in Figure 11-19.

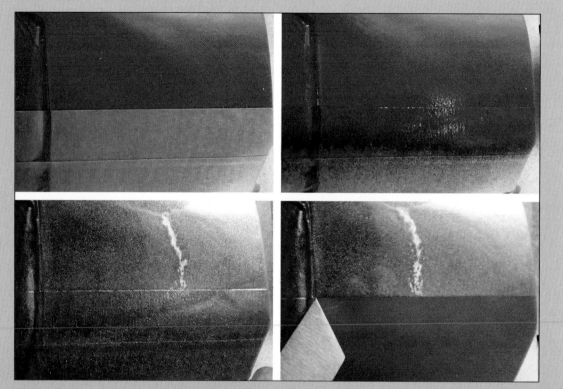

FIGURE 11-19: Painting along the edge of the tape with the basecoat color

This way, if any paint bleeds under the tape, it'll just be more basecoat. When it dries, it'll seal the gap. Then you can paint your secondary color over it. Remove the tape and you'll have an amazingly crisp line, as shown in Figure 11-20.

FIGURE 11-20: Sharp lines will make your parents proud, amaze your friends, and impress potential mates.

When selecting tape, be sure to get a width that makes sense for the project you're working on. If you're only masking off small areas, you don't want to use a roll of tape that's 4″ wide or you'll waste a lot of time cutting it into small pieces.

Applying Masking Tape

Tear a length of tape off of the roll and apply it to the surface. Press down the edges as you go. Don't stretch the tape, or it might lift or break during painting.

After you apply painter's tape, make sure to press down the edge to seal it. Otherwise, paint is sure to bleed under the edge of the tape and blur the nice, sharp line you're painting. Secure the tape to smooth surfaces by pressing the edges down with a putty knife, tongue depressor, or other flexible tool. Don't use tools to press the tape onto medium or heavy textures, because they cause the tape to tear.

Corners

To get a perfect fit on inside corners, start by running a piece of tape along the first edge. Leave the end a little long so that it goes past the corner. Press the tape down into the corner with a putty knife. Then cut along the crease with a sharp utility knife and remove the cut-off piece, as shown in Figure 11-21. Now you've got a nice, sharp corner and

you don't have to be quite so careful about getting the next piece of tape all the way into the corner.

Masking Large Areas

Proper application of masking tape takes time, and the tape can get expensive. So if you have a lot of area to cover, don't just use tape. Add paper to cover large areas, as shown in Figure 11-22.

While you can buy purpose-made masking paper in the same places where you buy your paint, it'll add to the cost of the paint job. Since you'll be making something a bit on the smaller side, you can probably get away with using leftover newspapers or printer paper. Just make sure whatever you use is thick enough to prevent the paint from seeping through to the work piece below.

Once you've covered all of the areas where you want the basecoat to show through, it's time to go ahead and spray on the next color. Once that dries, you can do more masking and painting on top of that.

If you're a procrastinator, rest assured that most good-quality painter's tape can be left in place for as long as two weeks without causing any problems for the finish under it. During this time, keep the work piece in a dry environment with a fairly stable temperature and away from direct sunlight. Sudden changes in temperature and humidity can cause the paint to warp or shrink, which can lead to peeling or lifting.

FIGURE 11-21: Cutting the tape after placing it against the corner makes for a nice, sharp corner.

FIGURE 11-22: Masking large areas with paper can save you a small fortune worth of tape.

FIGURE 11-23: Peeling masking tape correctly isn't rocket surgery, but it does take some know-how.

In any case, after the painting is done, it'll be time to get the tape off.

Removing the Tape

To avoid peeling the paint off, either pull the tape off immediately once you're done painting or wait at least overnight for the paint to dry completely. Be careful, though. Sometimes, paint that feels dry to the touch hasn't hardened and fully bonded to the part. There's still a chance that it will peel off when you pull the tape.

Taking the masking tape back off requires just a bit more skill than simply finding the edge and peeling it. You'll want to start by peeling it from one end at a moderate speed, pulling the tape at a 45° angle, as shown in Figure 11-23.

If the masking tape isn't removed right away while the paint is still wet, it may be a good idea to score the edge of the tape with a putty knife before you pull it off, as shown in Figure 11-24. This will break any bond that has formed with the paint and ensure that the masking tape will come off cleanly without damaging your paint job.

Shiny!

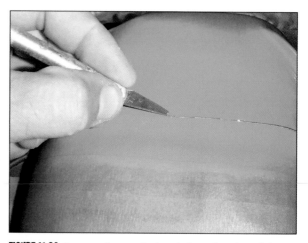

FIGURE 11-24: Once you've cut the bond along the edge of the tape, remove it as you normally would before you remove the tape.

So that's a quick overview of the basics of painting and picking out details. But if you're still feeling like your thing doesn't quite look realistic yet, don't despair. In the next two chapters, you'll learn new methods to really bring it to life!

12

FAUX FINISHES
Making Something into Something Else

FAUX = FAKE.

This whole book is basically about how to make things look like other things. If you've been following along, you've been building things out of scraps of plastic, paper, cardboard, and pipe, and random bits of pressed sawdust. While it's easy enough to just paint them a particular base color and leave it at that, sometimes it needs to be a bit more complicated. In this chapter, we're going to look at some creative ways to turn all of the random stuff you've been making things out of into believable pieces of metal, wood, and so on.

Paint Your Garbage!

Each of these techniques can take some time and experience to effectively master. It's always a good idea to practice on a piece of scrap material until you've got the hang of it. This also provides a good opportunity to test your coatings for compatibility and be absolutely certain that they won't cause problems when used together.

Metal

The creators of our favorite science fiction and fantasy characters kind of hate us. That's the only thing that would explain why their creations are usually clad in materials that retail for about the same price as a mid-size nation's annual defense budget. Since we're probably not working with that kind of cash, we'll have to fake it with some kind of paint.

Metal surfaces can be exceptionally tricky to simulate. In order to make metallic paint, tiny bits of metal powder (usually aluminum) are ground down and mixed with the paint. Adding different pigments will change the color and, from a distance, these bits of metal catch the light and look convincingly like a piece of metal. But upon closer inspection, something that's coated with gold metallic paint is usually never quite as shiny as actual gold.

In almost every case, metallic paints should be applied by spraying. Whether it's through an automotive spray gun, airbrush, or even the lowly aerosol can, when you're trying to get a bunch of tiny flakes of metal to lie down flat on the surface, there's usually no better way.

When it comes time to choose a metallic color, it's usually a good idea to pick something that's actually a shade or two brighter than the final desired finish. Adding some weathering as described in the next chapter will tone down the shine and make it darker (Figure 12-1).

That said, at some point you'll have to make something that calls for a shiny, polished finish with the look of chrome, gold, platinum, or whatever else that can't just be faked with a can of paint from the local hardware store.

FIGURE 12-1: Starting with bright silver paint (left), a little bit of weathering makes a pretty convincing steel plate color (right).

Chrome and Other Highly Polished Surfaces

While shopping at the hardware store for chrome paint, you may have seen a spray can with a bright, shiny, mirror-like cap and thought, *"Finally, the perfect chrome finish in a can! At long last, I can die in peace!"* Then you took it home and sprayed a coat onto something. That's when you found out that it was just ordinary silver-ish paint. The quest would go on, and life would continue to be a Sisyphean torment; searching forever for a believable chrome paint in a spray can.

That was then.

Nowadays there are a few options for believable chrome paints that can be used without specialized equipment. A few examples have brand names like Mirror Chrome or Looking Glass or MirraChrome or SuperChrome or Alclad. While most neighborhood hardware stores won't carry these paints, hobby stores and online retailers usually do.

These metallic paints all tend to have a few things in common:

- First, they all need to be sprayed over a perfectly glossy, smooth, black basecoat. Here is where proper planning and prior preparation pay off. Period.

- Second, they have to be applied in exceptionally light coats, allowed to dry, then coated again.

- Finally, being very delicate finishes, they all need to be protected with a gloss clearcoat if they're going to be handled at all.

Still, as long as the manufacturer's instructions are followed to the letter, these paints can reproduce remarkably shiny metallic surfaces, as shown in Figure 12-2.

FIGURE 12-2: So shiny! So chrome! So . . . painted?

Brushed Steel

Oftentimes, when a piece of steel is machined, or even if it's filed down by hand, the surface will have tiny parallel grooves scratched into the surface, leaving it with the "brushed" look shown in Figure 12-3.

An easy way to simulate this look is to start with a coat of black primer on your object. Then spray on a coat of silver paint. Before the silver paint has a chance to dry, gently drag the tips of a bristle brush over the surface to leave behind thin lines in the surface of the paint, as shown in Figure 12-4.

Be sure to keep the lines straight and parallel. As always, practice makes perfect. Figure 12-5 shows a good example of a piece with this brushed steel effect.

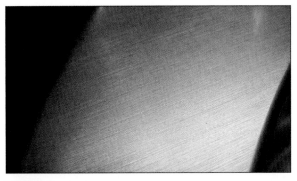

FIGURE 12-3: Brushed steel

FIGURE 12-4: Adding brush marks to the smooth coat of sprayed silver paint

FIGURE 12-5: Not really brushed steel

Another good option is to start with a glossy silver basecoat. Then, after the basecoat has thoroughly dried, apply a very light coat of watered-down black acrylic paint. Before the black paint has a chance to dry, use a sponge, rag, or a dry brush to remove most of the black using long, parallel strokes, as shown in Figure 12-6.

When you're happy with the fine streaks that have been left behind, allow the piece to dry, and you'll have a decent-looking piece of brushed metal, as shown in Figure 12-7.

FIGURE 12-6: Gently removing most, but not all, of the black paint with a rag leaves streaks behind.

FIGURE 12-7: A decent-looking piece of brushed metal, but made out of plastic

Leather

While there are a lot of great ways to simulate the look and texture of leather, you can get some of the best results by using a flat basecoat and then going back over it with a gloss topcoat. Here's what you'll need:

- The thing to be turned into "leather"
- A rich, flat brown for the basecoat
- A significantly different rich gloss brown for the topcoat
- Brushes or rollers to apply the paint
- Some plastic wrap

In this case, we'll be applying the faux leather to the handle of our unnamed* wolf warrior's battle axe. This technique could just as easily be used on flat or curved pieces.

Step 1

Paint the base color and allow it to dry completely (Figure 12-8).

FIGURE 12-8: The handle of the axe with the base color painted on

*What if we call her, *Mercilla the Mighty, Ruler of the Ruthless Raiders of Randelle*? Never mind. That's terrible.

Step 2

Use a brush or roller to apply a heavy coat of the gloss topcoat (Figure 12-9).

FIGURE 12-9: A coat of slightly different gloss brown covering the entire handle

🔫 Maker Tip

As with all of these techniques, it's always a good idea to test this technique on a scrap before you begin the finished project. It may turn out the colors look horrible together and changes are easier to make before the new paint job becomes permanent.

Step 3

Leave the paint to set for one or two minutes. Seriously. Stop touching it.

Step 4

Really, stop touching it. In fact, to keep your hands busy, spend this time by cutting a piece of plastic wrap into a rectangle that will cover the entire handle. Crumple the sheet into a ball. Then, stretch it out and lay it gently over the wet paint.

Step 5

Squish the wrinkly plastic into the wet paint. This will push all sorts of wrinkles into the paint (Figure 12-10).

Step 6

Once the entire surface is covered and wrinkled, carefully peel away the plastic sheet to reveal the design (Figure 12-11).

FIGURE 12-10: The wrinkling

FIGURE 12-11: Leather enough to fool most folks

If you want more of the basecoat color to show through, crumple up some more plastic wrap and repeat Steps 4–6 as many times as necessary to achieve the desired look.

If you want to make it more durable, once the second color coat has dried, the whole thing can be sealed with a matte or satin clearcoat. After sealing, a quick blackwash will also help to pick out the seam lines in between the pieces of faux leather (Figure 12-12).

FIGURE 12-12: The leather-wrapped handle of the battle axe

Wood Grain

There are a lot of really great ways to make things look like they're made of wood. So many in fact, that craftsmen skilled in the art of *Faux Bois* (French for "fake wood") can demand top dollar for making steel fire doors or sheetrock walls look like expensive wooden panels. While these folks can spend years mastering this craft, here are a couple of quick and easy options that will generate passable results. As always, feel free to experiment and combine these different techniques for better effect.

Woodgrain Option 1: Mostly Just a Bad Paint Job

If you're just trying to make something look like a tight-grained piece of hardwood, you're in luck. What would normally be a really poorly done paint job will actually get you halfway there.

Case in point: take a look at the handle of our unnamed* wolf warrior's axe. Figure 12-13 shows it with a coat of red primer (which is usually more of a chocolate or rust brown color).

FIGURE 12-13: Axe handle in brown primer

Using a cheap bristle brush and a darker (or lighter if preferred) brown paint or wood stain, paint can be brushed on, leaving very noticeable brush marks, as shown in Figure 12-14.

FIGURE 12-14: Visible brush strokes do a good job of simulating tight-knit wood grain similar to hardwood.

---

* What if we called her *Stephanie*? No? Okay.

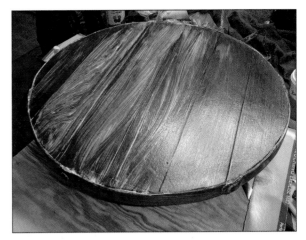

Maker Note

When using this method, it's important to make sure that the brush strokes aren't straight and that they run continuously for the entire length of the piece being painted.

FIGURE 12-16: The shield of . . . shall we call her *Farella the Fighter*? No . . .

Woodgrain Option 2: A Graining Tool—The Quick and Easy Route

While perusing the aisles at better-stocked hardware or paint stores, hyper-observant shoppers may have noticed something like the device shown in Figure 12-15.

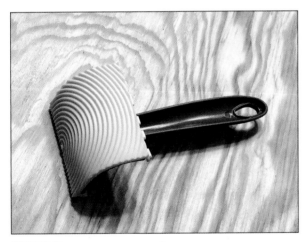

FIGURE 12-15: This is a graining tool.

Graining tools can be used to great effect when something fairly flat needs to look like wood. It just so happens that our unnamed wolf warrior woman will need a shield that will be big, flat, and wooden. But for now it's just made of EVA foam (Figure 12-16).

Before making something look like wood, it's a good idea to look at some examples. Check out the wooden furniture at home or go to the neighborhood home improvement store or lumber yard to look at examples of different types of wood grain. Determine which one you'd like your part to look like. If you can bring home a sample piece to match, that's even better.

After figuring out what wood should look like, here's what's needed to fake it:

- Basecoat paint matching the lightest shade of brown in the wood

- Gel wood stain that matches the darkest shade of brown in the wood

- Paintbrushes for the basecoat and wood stain

- Wood graining tool

- Whatever rags, water, or solvents that'll be needed for cleanup afterward

After gathering all of your tools and materials, it's time to begin.

Step 1

Prep and prime the part. In this case, the foam has been sealed with Mod Podge, but Plasti Dip rubberizing paint would work, as well.

Step 2

Apply a solid basecoat to the part with a paintbrush. Normally the goal is to get a smooth coat, but having visible brush strokes at this stage is actually a good thing. Try to have them run all the way along the length of the part, much like woodgrain would on a wooden part (Figure 12-17).

Step 3

Allow the basecoat to dry completely.

Step 4

Use masking tape to cover the edges of every other plank, as shown in Figure 12-18.

Step 5

Apply an even coat of the wood stain. In this case, we'll start with just one of the planks on the shield.

Step 6

While the stain is still wet, slide the wood grain tool along the length of the plank, as shown in Figure 12-19. Be sure to rock the tool back and forth while pulling it. This will create the irregular, uneven wood grain patterns.

Step 7

If the edges of the darker stained areas contrast too sharply with the lighter areas where the basecoat shows through, gently brush across them with a dry bristle brush, as shown in Figure 12-20. This will soften the edges and help to simulate the look of natural wood grain.

FIGURE 12-17: This sloppy basecoat, complete with visible brush marks, would normally be a problem.

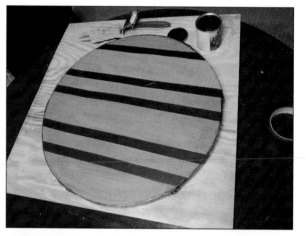

FIGURE 12-18: Masking on the edges of every other plank

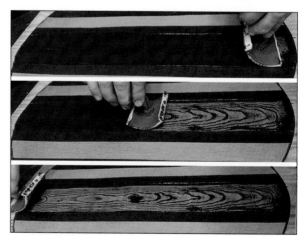

FIGURE 12-19: Wood graining is really just a matter of keeping everything random.

FIGURE 12-20: Dragging some of the stain with a bristle brush to soften the high contrast of the wood grain.

Step 8

Repeat Steps 4–8 for every other plank. On a piece like this, it's a good idea to first apply the stain and wood grain to alternating planks in order to keep them looking like distinctly different pieces of wood.

Step 9

With the first half of the planks grained, remove the masking tape and allow the surface to dry (Figure 12-21).

FIGURE 12-21: The first half of the planks dried.

Step 10

Repeat Steps 4–9 for the remaining planks.

🔫 Maker Tip

In addition to the curved area of the graining tool, other wood grain effects can be achieved by dragging the ends of the tool through the wet stain (Figure 12-22). Changing from one side to the other on alternating areas of the piece will add to the illusion that it's made from multiple pieces of wood.

FIGURE 12-22: The straight edges of the graining tool are also textured to simulate other wood grains.

Step 11

After all of the planks have been wood grained, set the whole thing someplace warm to dry.

Step 12

If you'd like to seal the surface, you can apply a clearcoat of polyurethane or Mod Podge using a paintbrush. Allow the finish to dry thoroughly, and then apply a second coat.

Step 13

Bask in the glory of what you have wrought (Figure 12-23).

FIGURE 12-23: The faux bois shield, defender of the mightiest of warriors

Carbon Fiber

Not everything is made of wood and metal. While most of your more modern plastics and composites end up just being solid colors because they lack any real aesthetic value, carbon fiber composites contain woven fibers that catch the light in all sorts of beautiful ways. This means that, unlike cheap plastic or pedestrian fiberglass, carbon fiber composites are often manufactured with clear resin surfaces that allow the weave of the fiber cloth to show through.

While it looks great, carbon fiber can be expensive. Since we're on the subject of faking things, here's what you'll need in order to give something plain and ordinary the striking look of a carbon composite.

To start with, you'll need the following tools and materials:

- Black spray paint
- One or two colors of silver or gunmetal gray spray paint
- Gloss clearcoat
- Rubber shelf liner or non-slip rug pad (Figure 12-24)

Step 1

Start with a black basecoat. This can be glossy or flat, and even subtly metallic if need be.

Step 2

Use the shelf liner as a masking template, as shown in Figure 12-25. It must be held tight against the surface.

FIGURE 12-24: The shelf liner has an ideal checkerboard pattern.

FIGURE 12-25: Shelf liner mesh taped in place as masking template

Step 3

Spray a series of *very light,* parallel, diagonal stripes of the silver or gunmetal onto the shelf liner.

Step 4

Remove the shelf liner and allow the paint to dry thoroughly (Figure 12-26).

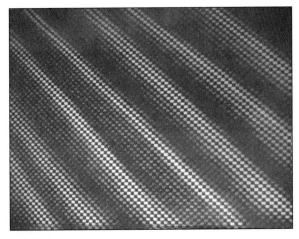

FIGURE 12-26: The striped checkerboard pattern after removing the shelf liner

Step 5

After the coating has thoroughly dried, apply a high-gloss clearcoat to really bring the finish to life, as shown in Figure 12-27.

This technique can be adjusted in a variety of ways to alter the final appearance and make it even more convincing.

First, after applying the metallic color and allowing it to dry, reposition the shelf liner with just a slight offset and apply a second metallic coat in a different shade. This will give the fake carbon fiber weave a bit of a "drop shadow" effect that will add depth to the texture and make it more believable.

Second, if you'd prefer to be able to see what you're doing while you're doing it, you can spray an even coat across the shelf liner instead of just spraying diagonal stripes. This will make a consistently bright checkered pattern on the surface. Once that dries, you can spray lightly misted black stripes across the surface and get the same effect with a bit more control.

Third, consider applying the lighter color first and then adding the darker color on top, or even starting with a silver basecoat and adding the checkered pattern in black.

Fourth, when insanity really sets in, experiment with different colors for basecoats, topcoats, and tinted clearcoats (also called candy coats) for even more unworldly results.

FIGURE 12-27: The glossy clearcoat helps bring out the composite appearance.

13

WEATHERING

The Fastest Way to Wear and Tear

NOW IT'S TIME TO know and love the most valuable word in this whole book: *verisimilitude*.

> *verisimilitude*
>
> *noun | ver-uh-si-**mil**-i-tood*
>
> 1. The quality of seeming real.
>
> 2. The semblance or appearance of truth; likelihood; probability:
>
>   *The story lacked verisimilitude.*
>
> 3. Something, as an assertion, with only the appearance of truth.

Verisimilitude is where a prop maker really proves their worth. Sure, maybe you can take all sorts of oddball things and make them into whatever it is you're trying to make. You add a few coats of paint to hide the fact that the pieces are made out of wood and plastic and paper and auto body filler and garbage, but everything still looks pretty fake (Figure 13-1).

In order to add the element of verisimilitude, you must embrace the general state of filth that is reality.

FIGURE 13-1: Still seems like a plastic toy

This is where weathering comes in.

No matter how clean something is, no matter how well it's maintained, and no matter how much someone cares, if it exists in the real world, it'll show some dirt (Figure 13-2). On top of that, for some reason, everything looks cleaner and shinier in photos and on film than it usually does in real life. This means that if you want something plastic to start looking real, you're going to have to add some ugly to it.

FIGURE 13-2: Even though someone probably loves this car dearly and washes it often, even a few minutes on the road leaves all kinds of dust and grime behind.

So. Remember that nice, shiny paint job you worked so hard to get just right? Now's the time to mess it up.

Additive Weathering

When it comes to weathering, there are two basic routes. You can either add more paint, or find ways to take paint away. The easiest is the additive process. Starting with that nice, clean finish, you'll be deliberately adding simulated dirt, dust, and scuff marks wherever you want them. The only real challenge is making sure it doesn't look deliberate.

Washes

If you do nothing else at all, add a wash. This will do a great job of helping all of the little details and recesses really stand out. It'll help the finished piece look good in almost any lighting, and ensure that all of the edges, seams, and recessed details you painstakingly replicated are still visible even with a harsh camera flash or bright outdoor sunlight.

There are two ways to go about this. First, start with some black acrylic paint from your local craft shop or art supply store. Mix some of the paint with some water—lots of water. Then, brush a very wet coat onto the entire piece to be weathered, as shown in Figure 13-3.

FIGURE 13-3: Everything wet

🔫 Maker Note

This is a wet, messy process. Find a place to work where that won't be a problem, and lay down some newspaper or a drop cloth to make cleanup easier.

The watered-down paint will find its way into every nook and cranny. Let the darkness flow over everything. Take care to ensure that it doesn't form any pools or puddles. The idea is to darken the deeper parts of the piece in order to increase the contrast and pop the details out a bit.

Now let it sit and dry. That's it. Don't help it along by dabbing off some of the wetness from the high points with a paper towel. Don't try to dry it off with a hair dryer. Don't blow on it. Just leave it alone. Sit it in the sun or under an incandescent light if you must. In the end, you'll have a very thin layer of darkness covering the entire thing, with even more darkness concentrated in all of the recessed details on the surface (Figure 13-4).

The other option is a much more deliberate process and will result in a significantly heavier layer of grime. Here's what you'll need:

- Black acrylic paint
- Isopropyl alcohol
- A cheap bristle brush
- A cup

- The most expensive imported bottled water you can find (or just tap water)
- Rags or heavy-duty paper towels

Start by taking the un-thinned acrylic paint and spreading some of it onto the surface of the part, as shown in Figure 13-5.

Dip the brush into a cup of water and brush the black acrylic goop over the entire thing (Figure 13-6).

Continue adding water and smearing the paint around. Along the way, the acrylic paint will be thinned slightly. You may notice that the acrylic paint will bead up and leave little bare spots where it doesn't want to flow, as shown in Figure 13-7.

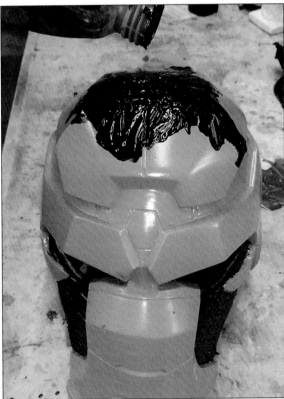

FIGURE 13-5: Spreading some goop

FIGURE 13-4: Layer of darkness? Check.

FIGURE 13-6: More goop

FIGURE 13-8: Take that, surface tension!

FIGURE 13-7: Surface tension can be your enemy.

The best way around this is to pour a bit of isopropyl alcohol onto your brush and smear it into the acrylic as you spread it around. This will break up the surface tension and add a bit more randomness to the finish (Figure 13-8).

Once everything is completely besmirched with black acrylic goop, take a wet rag or paper towel and start dabbing away at the surface (Figure 13-9) in order to thin out the paint on the high points of the piece.

FIGURE 13-9: Making the darkness a bit lighter

After wetting everything down and smearing the black goop with the wet rag, switch to a dry rag or paper towel and start removing the black paint from the high points. This will leave behind a grimy blackness that will fill the low points but also leave a few smears on the high points to give the whole thing a thoroughly used look, as shown in Figure 13-10.

Drybrushing

Look around any place where tools or weapons are used and there will be countless signs of wear and tear. No matter how well someone cares for a piece of hardware, eventually the corners will be worn from use, the paint will be scratched, or the aging metal will have parts that are polished from use and handling, as shown in Figure 13-11.

The best way to get these kinds of wear marks and worn edges in a hurry is called *drybrushing*. To start, gather up the following materials:

- An old paintbrush
- Metallic paint
- Black paint
- A rag or paper towel

When drybrushing, the object of the game is to get just a bit of paint to adhere to the sharper corners and raised edges. Begin by dipping just the very tip of the brush into the paint, then wipe most of the paint off of the brush onto the rag or paper towel (Figure 13-12).

Once the brush is nearly free of paint, brush it across the edges where the wear should be (Figure 13-13).

FIGURE 13-11: The edge of a toolbox that has been places and seen things you could never imagine

FIGURE 13-10: Dirty, oily, grimy used look. Done.

FIGURE 13-12: Wiping most of the paint off the brush

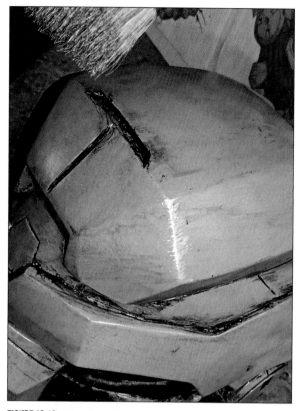

FIGURE 13-13: After dragging the nearly dry brush across the part, paint is only left on the raised edges.

FIGURE 13-14: Hunter's used helmet

Repeat as necessary until the ideal look has been achieved. Figure 13-14 shows a copy of Hunter's helmet with some subtle wear marks along the edges.

Metallic wear marks on a piece that's a very light color may not stand out as much as you'd like. In this case, it's often a good idea to start by drybrushing with black paint first, then go back over those same areas with the metallic paint, leaving just a bit of black showing around the edges. Figure 13-15 shows Hunter's helmet with black scratches drybrushed on.

Once the black paint has dried, go over the same areas with the metallic paint. Be sure to leave some black paint visible at the edges (Figure 13-16). These outlines will make the bright metallic areas stand out better against the lighter paint color.

FIGURE 13-15: Black scratches

FIGURE 13-16: Shiny scratches and wear marks outlined with black for contrast

Subtractive Weathering

Drybrushing some wear marks onto the surface works out pretty well for areas where the paint will likely be worn away when something is being used, but when you want to show paint that will look like it's been chipped or scratched when someone takes a closer look, it's time to step up your game a bit more.

The process of subtractive weathering involves a bit more planning and a lot more time. Start by gathering the following tools and materials:

- Shiny metallic paint
- Clearcoat
- Some color of paint that will pass for primer
- Some color of paint that's a tiny bit darker than the topcoat color
- The paint to be used as a topcoat
- Brushes

- Toothpicks, popsicle sticks, or some other thin or pointy bits of wood
- A liquid masking agent (mustard, rubber cement, or toothpaste will work)
- A rag or paper towels

Once the piece is prepped and primed, start with a glossy coat of silver paint (Figure 13-17).

FIGURE 13-17: Shiny

Chrome paints are often somewhat delicate, so after allowing plenty of time for the paint to cure, it may be a good idea to add a few coats of clearcoat. Once the clearcoat is dry, pick out all of the areas where scratches will show through to the bare metal and cover them with the masking agent. In Figure 13-18, these areas are covered with mustard.

FIGURE 13-18: Blobs of mustard, a common condiment also known as, "Ew, gross; how long has this been in your shop?"

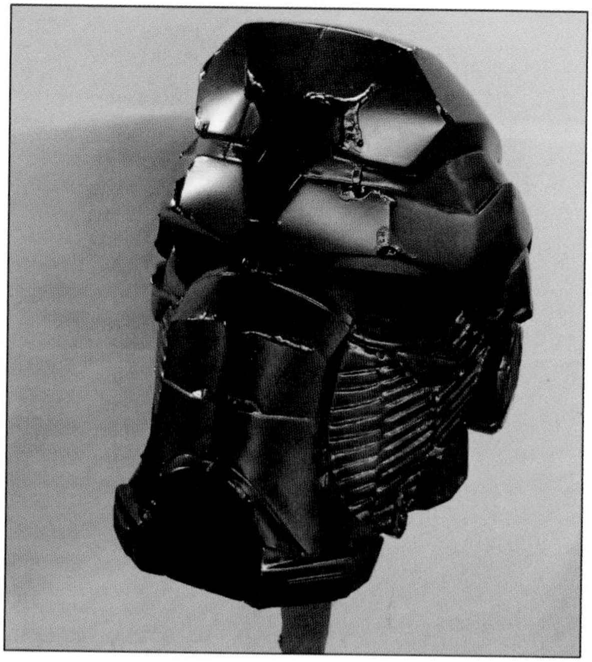

FIGURE 13-19: Primer color

🔫 Maker Note

Mustard is an ideal masking agent. This was determined through a months-long series of clinical trials to verify it has just the exact pH balance to counteract the corrosive activity caused by the solvents in the spray paint. Actually, no that's not true. In reality, this same masking effect can be achieved with anything viscous enough to stay put once applied. Ketchup will probably work. Salsa verde will probably not. The point is to use something water-based that the oil-based paint can't stick to.

With the masking in place, give the whole thing a coat of whatever primer color has been selected—black in this case (Figure 13-19).

When the primer color has dried, you'll notice blisters full of masking agent trapped under the paint. LEAVE THEM THERE. If these blisters were scratched off at this point, they'd leave behind areas of shiny metallic basecoat showing through the primer layer. Instead, add just a bit more masking randomly around the edges of the blisters. While you're at it, add a few new masked areas, as well (Figure 13-20).

These additional masked areas will leave behind areas where the topcoat will show through to the primer layer, but not to the bare metal area. Now it's time to spray on a coat of paint that's just a bit darker than the desired topcoat color, as shown in Figure 13-21.

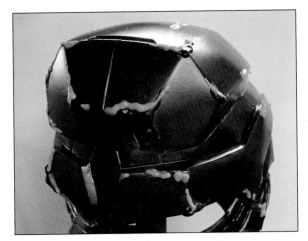

FIGURE 13-20: More masking

FIGURE 13-22: Even more masking

FIGURE 13-21: Just a bit darker than the desired topcoat

Once that layer dries, it's time for just a bit more masking (Figure 13-22). Areas where the topcoat will be scratched will show through to a slightly darker color, but not all the way down to the primer or the bare metal.

Now spray on the final topcoat color, as shown in Figure 13-23.

FIGURE 13-23: The final topcoat

After the whole thing has dried, use a damp rag or paper towel to pop open the blisters of masking agent and gently scrub them off. At this point, additional scratches can be added by carving into the paint with a toothpick or other semi-sharp wooden implement (or even a fingernail), as shown in Figure 13-24 and Figure 13.25.

FIGURE 13-24: Adding more scratches to the edges of the masked areas

FIGURE 13-25: This shows the Hunter helmet with scratches and paint chips added using this subtractive technique.

With all of that work done, a blackwash will still do a great job of getting all of the details to stand out and really bring the whole thing to life (Figure 13-26).

FIGURE 13-26: Again with the blackwash

Mud

When the paladin warrior dons his highly polished armor at the beginning of a tournament at the castle, it may be all shiny and new-looking at first. But after just a few minutes into the day's battle, even the most persnickety person on the field is going to end up with dirty boots at the very least. Now imagine what the warrior would look like after a forced march across a hundred leagues of forest trails or after sneaking behind the enemy flank for a week. There's no way the armor will still look clean after that kind of abuse.

While it's a good idea to add a bunch of general dirt and grime to help pick out the details and make things look used, special care and attention need to be taken around the feet or anywhere else that is likely to have been stuck in the mud (i.e., knees, elbows, and so on).

Simulating a light layer of dust, mud, or dirt can be done by haphazardly spraying the areas where dirty is desired with various shades of brown, tan, and black spray paint. Holding the can a little bit too far away from the surface being painted will cause the paint to form small droplets as it coalesces on the way through the air (Figure 13-27). This helps with the illusion of specks and flecks of mud and dirt sticking to the surface.

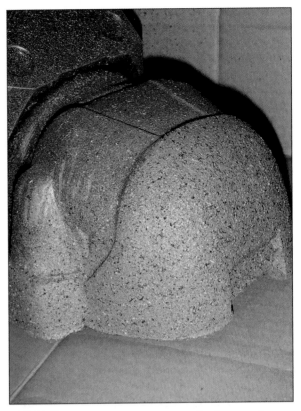

FIGURE 13-28: A stone-textured paint dusted with two or three other shades of brown and black make for the ultimate muddy shoes.

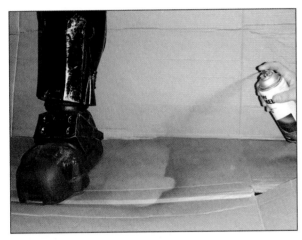

FIGURE 13-27: Holding the spray can too far away from the surface will cause paint droplets to form in the spray pattern.

If an even heavier layer of caked-on dirt is desired, you'll want to switch to any one of a number of textured spray paints available in most hardware stores. The thickest, dirtiest versions are actually two-tone paints intended to simulate the look of a stone surface. Adding one or two of these paints into the mix will really help make the dirty parts look even dirtier (Figure 13-28).

Rust

Every person since before the dawn of the Iron Age knows what rust looks like. Or so they think. The first instinct is just to think it's the reddish-brown color that covers old metal stuff. Upon closer inspection, rust can have a lot of personality. Think it's just brown? Wrong! Heavy rust can take on hues of purple, yellow, and even hints of white. It can form in blisters in deep corners or recesses between parts, or it can be just a light coating over the top of exposed steel.

Before you add rust to a project, give it some thought. Is this a newly forged piece of steel that

FIGURE 13-29: Rust spotted in the wild

was carried on the long march to the battlefield without oil or polish for a couple of days? Or is it supposed to look like something that was found at the beach after a year in the salty air? Either way, adding rust calls for a bit more work and thought than many people might presume. Start by going outside and studying some real-word samples of rust. Figure 13-29 shows just a few examples of the whole wide world of rust variations.

There are lots of different products you can use to replicate a rust finish, along with different techniques you can choose. To simulate the look of light surface rust on a piece of steel that's been left bare and unpolished for a few days, try brushing on some very watered-down acrylic paint in a reddish-brown color. The "rust" will pool in the recesses, while still leaving a bit of color on the outside edges.

Heavy Rust

Before faking some really heavy rust, gather up the following tools and materials:

- Red primer
- Acrylic paints in a selection of rusty colors like burnt sienna, mustard yellow, white, purple, and terra-cotta
- Sea sponge(s)
- Spray clearcoat in flat or matte
- A bowl or cup with some clean water
- Baby powder or talcum powder

Step 1

Start with a nice coat of red primer, as shown in Figure 13-30.

FIGURE 13-30: Red primer already starts making things look rusty.

Step 2

Dampen the sea sponge with water and use it to daub on a blotchy coat of terra-cotta paint (Figure 13-31).

FIGURE 13-31: The basecoat in terra-cotta, all blotchy and uneven

Step 3

Add some randomly scattered blotches of the other shades of brown wherever the rust is supposed to be (Figure 13-32).

FIGURE 13-32: More multicolored blotches of reddish to yellowish brown

Step 4

Apply even fewer highlights of off-white, yellow, and purple. These highlights of white and yellow add dimension to the finish so it doesn't have a "flat" appearance, but use them sparingly (Figure 13-33).

FIGURE 13-33: Just a few blots of off-white, yellow, and purple

Maker Note

Applying paint with a damp sponge can leave repetitive, recognizably artificial markings. To avoid this, change your grip from time to time. Be sure to use all sides of the sponge in order to make the rust surface appear random.

Step 5

Use some burnt sienna over the top of the purple, yellow, and white areas to blend them back into the rest of the rust colors and tone them down a bit. The highlights should remain visible, without standing out against the rest of the rust.

Step 6

Allow the acrylic paints to dry.

Step 7

Seal with a matte sealer. This won't just protect the surface from wear and tear, it will also help to bring out the texture of the rust blotches. Just as importantly, if there's any kind of gloss sheen to the acrylic paints that were used, you'll want to cover that up with something that is not glossy at all.

Step 8

While the sealer is still wet, dust it lightly with baby powder or talcum powder, as shown in Figure 13-34. The object is to add a bit more texture to the surface, but don't overdo it. As the sealer soaks into the powder, it should still be transparent.

Once the sealer has dried, you'll have a believably rusty surface, as shown in Figure 13-35.

Step 9

Blackwash the piece to add even more depth to the finish (Figure 13-36).

If you'd like to bring back some of the look of bare steel, you can always drybrush some silver onto the edges where the metal would be worn clean, as shown in Figure 13-37 on the next page.

Light Rust

Suppose that, instead, you need to make your bare metal parts look used, but not completely neglected. In this case, you'll want to start with a silver basecoat, as shown in Figure 13-38.

In order to tone down the fake-looking silver paint, apply a blackwash over the entire thing (Figure 13-39).

FIGURE 13-34: Adding a bit more surface texture to the wet sealer

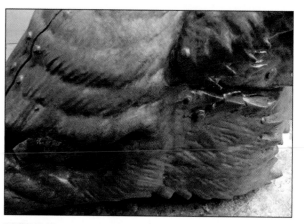

FIGURE 13-35: Rust (but not really)

FIGURE 13-36: Blackwash helps the details stand out.

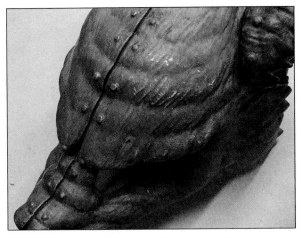

FIGURE 13-37: A bit of drybrushed silver paint on the edges makes it clear that this was once a shiny piece of metal.

FIGURE 13-38: Bare silver start for a less-rusted steel helmet

FIGURE 13-39: The blackwashed silver helmet looks a lot more like forged steel.

Next, thin down some brownish-orange acrylic paint and apply a very light wash over the whole thing (Figure 13-40). This will simulate the kind of light surface rust that covers bare steel after a short exposure to the elements.

Since it's harder to polish the rust out of tight corners, go back and apply a bit heavier wash to the recessed areas (Figure 13-41).

Now you've got yourself an aged steel helmet. But what if it was bronze, copper, or brass?

FIGURE 13-40: Light surface rust simulated by a wash with watered-down, burnt-orange acrylic paint

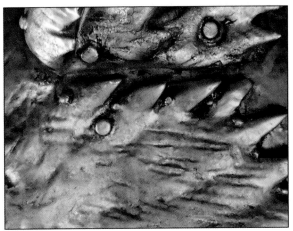

FIGURE 13-41: More rust in the corners

The Path to Patina

Not all metal will rust. In fact, those orange, red, yellow, and purple hues are unique to alloys that contain at least a little bit of iron or steel. Things made of copper, bronze, or brass will still oxidize, but they never rust. Instead, they morph into a completely different collection of colors referred to as *patina* or *verdigris* (Figure 13-42).

These layers of rich blues and greens come from years of aging. But if you have a convention or costume party rapidly approaching, there's no time to wait for it to happen naturally. Once again, it's time to fake it.

A believable patina starts with a nice, smooth metallic basecoat. In this case, we'll start with a copy of the helmet for our wolf warrior woman*, painted with an antique bronze paint (Figure 13-43).

One the basecoat is dried, the next step is to pick out some acrylic paints in a variety of shades of green (Figure 13-44).

The application process will be much the same as the rust process. The paint can be smudged on with a sponge brush, bristle brush, or sea sponge, but it's a good idea to leave raised areas of the basecoat exposed. Any place that's likely to have something rubbing up against it is going to be clean and polished.

Once the paint has dried, we will have a more appropriately ancient bronze helmet (Figure 13-45).

---

*We could call her *Romula Wolfemadchen* . . . Nah.

FIGURE 13-42: A strip of bronze after years in a marine environment

FIGURE 13-43: The beginning of an ancient bronze helmet

FIGURE 13-44: Patina colors range from teal to green to aquamarine with the occasional bit of white.

FIGURE 13-45: The ancient bronze wolf helmet

FIGURE 13-46: A lighter application of patina

Of course, maybe you don't want something that looks like a thousand-year-old piece of bronze. If you're aiming for something that looks like it gets regular use, but still has a touch of patina, the same process can be used in moderation. Figure 13-46 shows another casting of the Wolf helmet painted with a bright gold color, blackwashed to add a bit of age, then given a layer of patina into the deeper recesses.

After daubing off most of the patina with a damp rag and allowing it to dry, the end result is a much less aged brass helmet, shown in Figure 13-47.

Scorches and Burn Marks

The safest way to simulate this kind of damage is by using grimy, oily shades of black and dark brown applied via airbrush or carefully using a spray can. The key thing is to determine exactly how the marks were supposed to have been made. Was it a glancing blow from a bullet that got white-hot as it slid past the surface? Then the leading edge should be a bit sharper, while the trailing edge gradually fades, as in Figure 13-48.

FIGURE 13-47: A slightly old brass helmet

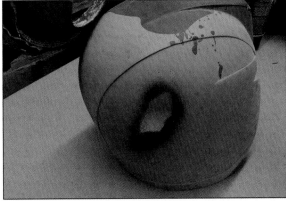

FIGURE 13-48: A little love with the airbrush simulates a blow from a laser blast that burned the paint and bared some of the metal below.

Was it caused by an explosion nearby that briefly engulfed the whole thing in a broiling, sooty blast cloud? Then a more generalized spraying of smoky shades of black and dark gray might make more sense.

The less-safe way to simulate these kinds of markings is to use FIRE. Once the paint job is otherwise complete, spray on a final clearcoat to seal everything. Then twist up a few pieces of electrical tape, hold them on one end with a pair of pliers, and light the other end on fire. The burning plastic tape will make a particularly heavy, oily smoke. While the clearcoat is still wet, hold the painted piece well above the flame so that the smoke leaves soot marks in the wet clearcoat, as shown in Figure 13-49.

When the clearcoat has dried, you'll have some very believable, random-looking scorch marks (Figure 13-50).

Burnt Metal

Suppose you've built yourself a lifelike replica of a deadly plasma rifle, laser blaster, or some other thing that gets really hot in daily use. Maybe your steam-driven power claw has an exhaust stack that really gets cooking when the claw is clawing. When metal bits are routinely exposed to high heat, they tend to change color in spectacular and permanent ways.

Before you try to simulate this, you'll want to take a look at some examples in the real world, like the one in Figure 13-51.

FIGURE 13-49: Sooty smoke sticking to the wet clearcoat

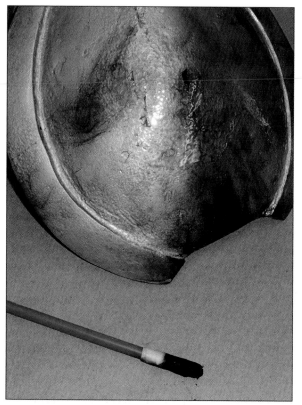

FIGURE 13-50: Sometimes the best way to fake it is to make it real.

🔫 Maker Note

Make sure the part is held far enough above the flame so that only the smoke is coming into contact with it. Too close, and you run a very real risk of having the whole thing catch fire. That's no kind of good time at all.

FIGURE 13-51: The heat from welding these two pieces of steel together causes a rainbow of tempering along the edges of the seams.

FIGURE 13-52: Sooty burnt muzzle on Hunter's rifle simulated with some flat black paint

The simplest version of burnt metal can be achieved by dusting the charred area with a bit of flat, black paint (Figure 13-52).

For a piece of metal that has been exposed to higher heat, things start to get a bit more interesting. Figure 13-53 shows the same burnt muzzle with a bit of gold paint sprayed on before covering most of it with flat black.

For even more interesting heat-treated metallic looks, it can be a good idea to use Candy Coat (a type of tinted clearcoat) and an airbrush. Figure 13-54 shows yet another resin cast of Hunter's rifle with a variety of blue, gold, purple, and brown candy coats blending into the blackened, burnt end of the muzzle.

Blood Spatter

If you've ever murdered someone with an axe, you may have noticed that the blood gets everywhere. Since it's actually a bad idea to murder someone with a fake axe, we'll have to settle for using paint to add a tiny bit of gore to blunt or edged weapons, as well as the resultant spattering on nearby armor parts.

FIGURE 13-53: A bit of gold (left), mostly covered with black (right), makes for an even more believable bit of heat-treated metal.

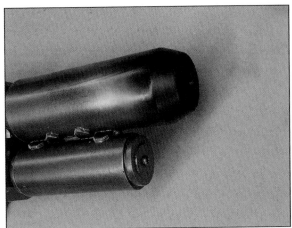

FIGURE 13-54: Much like the heat-tempered weld joints of a show car's exhaust manifold, Hunter's muzzle is kinda pretty.

Now any fool can spatter paint on something and call it bloody. Making it convincing requires a closer look at how blood behaves.

First off, it's not always red. Sure, it may start out bright red, wet, and shiny, but before too long, it will start to dry. As it does, it will darken and congeal. Eventually, it will turn into a brown crust, and finally it will become black and look nothing at all like blood to the layman.

This means it's a good idea to have a story in mind when adding blood effects. Was that mace just used to cave in a skull moments ago? That means it probably needs to have some pretty wet-looking, red blood on it. Is the blood spattered across those pauldrons a few hours old? Then maybe it needs to be a bit darker and thicker. Is today's killing just the latest in a lifelong career of carnage? If so, it's a good idea to mix it up.

The pattern of blood spatter will also tell a story. This is why blood-spatter analysis is such an important part of the forensic investigation of crime scenes. The size and shape of droplets can tell you a lot about the direction and speed that they were traveling right before they impacted whatever surface they're on.

Maker Note

Like most of the weathering processes, all of these methods will be messy. Wear gloves and cover your work area in order to minimize cleanup afterward.

This can be simulated in four basic ways: dripping, whipping, flicking, and smearing*.

_____

*If this reminds you of someone, we don't want to hear about it.

For discrete drops, start by thinning down your paint. Don't overdo it, though. As the saying goes, blood is thicker than water. It's thinner than mayonnaise, though. In any case, once the paint is thinned down to the point where it will flow properly, load up a paintbrush with a little more paint than you'd use for normal painting, then hold it over the area where the blood should go and let 'er drip (Figure 13-55).

The farther the paint drops, the bigger the splat will be, and the more interesting the outside edges will become. If the drop strikes the surface at an angle, that splat will stretch along the direction of travel, as shown in Figure 13-56.

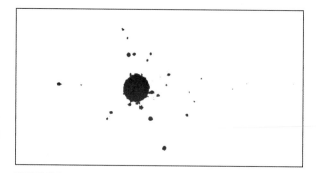

FIGURE 13-55: A single, discrete drop of shiny new "blood"

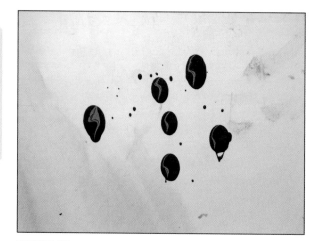

FIGURE 13-56: Oblong drops created by dripping paint onto an inclined surface

For a more dynamic spatter effect, load up the brush with even more paint, then snap it forward like a bullwhip. Momentum will carry the paint forward and cause streaks to lay across the surface being painted, as shown in Figure 13-57.

But what if you want a misty version of spatter? Say your character was standing behind someone who was shot and got covered in the sprayed mist. This can be simulated by taking a stiff-bristled paintbrush or an old toothbrush, getting paint on the tips of the bristles, then flicking your fingertip across the tips of the bristles. The paint will splatter all over the place and make a nicely random effect, as shown in Figure 13-58.

Now let's say you're dressing as a berserker, brawler, or otherwise handy kind of fighter who likes to keep their enemies within bayonet range. There may come a time when, in the last moments before counting coup and taking trophies, a not-dead-yet opponent reaches up and grasps, pushes, or otherwise flails against their killer. Along the way, blood gets smeared around. To simulate this, simply dip your gloved fingertips or hand into the wet paint and daub, slap, or otherwise smear it on (Figure 13-59).

After mastering these four basic methods, it's a good idea to combine them in some way (Figure 13-60).

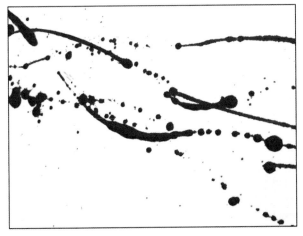

FIGURE 13-57: Streaks of blood spatter created by the whipping motion of the brush toward the surface

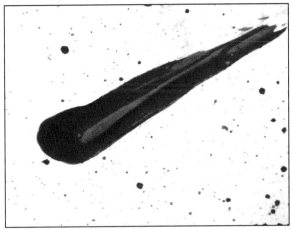

FIGURE 13-59: Fingerpainting with blood spatter

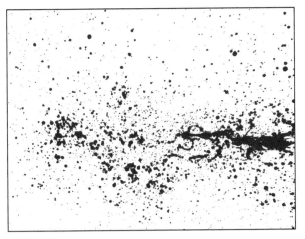

FIGURE 13-58: Flicking the paint makes great spatter effects.

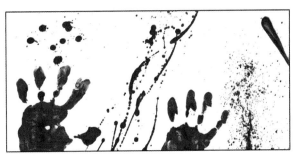

FIGURE 13-60: Combining methods to achieve a horror show look

🔫 Maker Tip

A little can go a long way. Unless your character just happened to be downwind of a suicidal grenade-swallower, there doesn't need to be all that much blood.

Of course, all of these examples just show fresh-looking blood. In order to make the gore look a bit older, start with splattering a dark brown color, like raw umber or burnt sienna, with a flat sheen, as shown in Figure 13-61.

Bear in mind that the thicker spots of the blood spatter will dry slower and stay red longer. So once the brown paint has dried, add a bit of gloss red to areas where the thicker blood hasn't quite dried yet

Finally, to add depth and wetness, a glossy clearcoat can be drybrushed on over the red parts (Figure 13-62).

FIGURE 13-61: Raw umber looks great as dried, scab-looking blood clots.

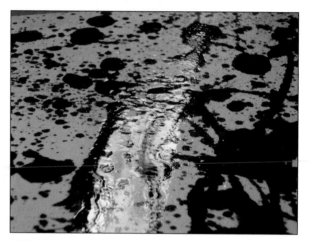

FIGURE 13-62: The red, red krovvy flows . . .

PART V

Bringing It All Together

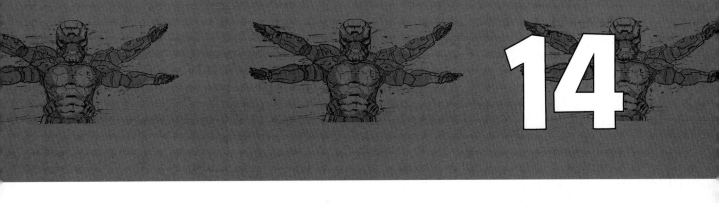

STRAPPING AND PADDING

Hold Yourself Together!

NOW THAT YOU'VE MADE all of the pieces and parts of your armor, it's time to figure out how you're going to wear it. After all, an awesome costume starts looking a lot less cool when pieces of it start falling off and you can't move in it.

The actual method used to hold a suit of armor together will vary greatly depending on the design of the suit itself. But no matter what shape the armor is, the object at this point is to maximize range of motion, as shown in Figure 14-1.

When you're in costume, you may not need the ability to run at a full sprint, dodge bullets, parry melee attacks, vault over low obstacles, and do a combat roll into a crouched fighting position, but if your armor is unwieldy and inflexible you're probably not going to have much fun wearing it. With that in mind, let's take a look at two absolutely vital elements of a suit of costume armor: padding and strapping.

Since everybody's body is at least a little different and a lot of the details will vary depending on the design of your armor, there's no best way to strap everything together. So this chapter will

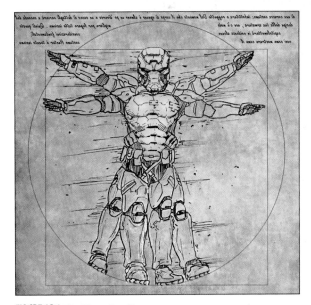

FIGURE 14-1: The Vitruvian Hunter

highlight a few key considerations, as well as show how the Hunter costume is held together.

Padding

If your armor is made of something heavy or rigid, common sense will tell you that it's a good idea to pad the inside in order to make it more comfortable. But even if it's made out of something light and flexible, adding padding will allow you to wear your armor comfortably without having it bounce all over the place. In addition to helping you avoid pinching and bruising, adding padding will keep everything snug and fitted. It'll keep the thigh plates and shin plates in place so they won't spin around, end up backward, and have you hobbling around the neighborhood on Halloween. It'll also keep forearm pieces from sliding around and grinding into elbow pieces and scraping off the paint. Finally, if you're not quite as bulky as the character you're trying to replicate, it can add some much-needed mass so that you properly fill out the costume.

When you're adding padding, work incrementally, building up a few sections at a time and

testing the fit along the way until you feel comfortable that the piece is staying where you want it.

Padding Materials

There are two readily available options for costume armor padding: EVA foam and upholstery foam (Figure 14-2).

EVA foam is the same foam as the floor mats that were used to make armor in Chapter 3. This is a good, firm option that will allow you to spread the pain of a heavy piece of armor rubbing against your person.

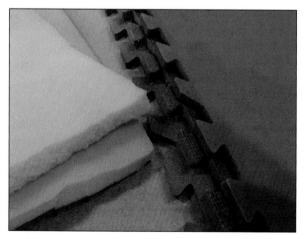

FIGURE 14-2: EVA Foam (right) and upholstery foam (left) are easy to find and easy to use.

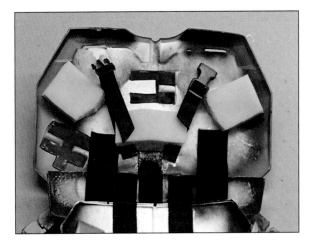

FIGURE 14-3: Gluing foam into Hunter's chest

Upholstery foam is the green or yellow stuff you can pick up by the yard at your local fabric store. It's available in various thicknesses and is much softer and lighter than EVA foam. It's also typically more expensive than the foam floor mats. If you're looking to pad something lightweight or add a lot of bulk without adding a lot of weight, upholstery foam is a great option.

Installing Padding

Once you've decided where to add padding, you'll want to attach it as permanently as possible. One of the best options for this is a spray-on contact adhesive called Super 77 made by 3M. To use it, you simply spray the two surfaces you want to adhere together, allow them to air dry until they are both tacky, and stick them together. Figure 14-3 shows upholstery foam blocks being installed inside Hunter's chest armor.

If you want a more temporary option, you can always use some black gaffer's tape (or even black duct tape) to stick some foam in place inside the armor. If you decide to make changes later, you can simply peel it off and reposition the pads.

Armor Attachment Options

When choosing a method to attach your armor to your carcass, there are a lot of considerations. Is the armor made of lightweight vacformed or foam pieces? Is it a bunch of 10-pound slabs of Rondo-reinforced Pepakura? Is there a particular set of clothing that will always be worn underneath the armor? Does the design include visible straps, or will they have to be hidden somehow? All of these factors will come into play in the final design of the attachment system.

Here are just a few options, as well as their pros and cons . . .

Velcro to an Undersuit

If the armor is lightweight, you might just be able to get away with gluing Velcro patches to the inside of the armor pieces, then sewing matching patches of Velcro onto the undersuit.

This can be a great option because it'll be easy to get the armor on and off in a hurry. The biggest drawback is that even the most reliable industrial-strength Velcro isn't able to reliably support very much weight once the pieces start moving as you walk around. You'll also have to make sure that your undersuit is made out of something that doesn't stretch too much, or the armor will start to sag.

The main concern with using Velcro sewn directly onto your undersuit is that you are now stuck using that base garment for this suit of armor and only this suit of armor. If your armor is made of foam, there's also a chance that the constant peeling action of removing the Velcro will start to tear up the foam armor from the inside.

Snaps to an Undersuit

Much more reliable than Velcro, snaps are great for holding your armor onto an undersuit. The challenge is finding a good way to attach the snaps to the armor parts. Gluing the snaps in place may seem like a good idea, but sooner or later the glue bond will fail and your pauldrons will fall off and go clattering across the floor.

If the armor is made out of sheet plastic, you can get away with mounting a snap to a piece of scrap plastic sheet, then solvent welding the scrap into place on the inside of the armor, as shown in Figure 14-4. This allows you to mechanically attach the snap, and then chemically bond it to the rest of the plastic.

If your armor design has places where you can get away with having practical nuts and bolts or rivets or screws visible on the surface, you can take advantage of them. Just be sure that your fasteners don't interfere with the functionality of the snaps. If your armor is made from something thick and rigid, you can run a rivet through the snap right into the piece to be attached and it'll hold itself in place.

Strap Harness So the Armor Holds Itself Together

If the costume you're trying to match has visible straps and buckles as a part of the design, life is easy. If not, you can still get away with just having the armor strap itself together. All you'll have to do

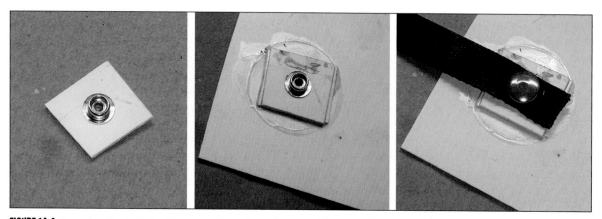

FIGURE 14-4: A good method of attaching snaps to vacformed armor without a visible attachment on the surface

is work out a good way to hide the straps or make them blend in.

Strapping Materials

If you do a bit of shopping, you can find countless options for costume straps. Here are just a few of the things you're most likely going to need:

NYLON WEBBING   One of the most commonly available strapping materials in the world today, nylon webbing is strong and inexpensive, and it's manufactured in a wide array of widths, thicknesses, and colors.

LEATHER   If you're going to have visible straps that need to look like they're made of leather, there's no substitute for the real thing. While it can be had in small quantities at some arts and crafts stores, bigger pieces can be found at saddle shops or tanneries for better prices.

FABRIC   If you need an odd shape and you don't want it to be made out of leather, you may be just as well off using hemmed strips of any of the countless varieties of cloth available at your local fabric shop.

COTTON AND CANVAS WEBBING   A natural alternative to nylon, it's a little harder to find, but it comes in a wide array of colors.

ELASTIC   While all of the above are great options for straps that don't stretch, sometimes you need something that will have a little bit of give. Elastic can be found in all sorts of widths and colors, and there are light- and heavy-duty options depending on how much stretch you need it to have.

VELCRO   Invented in the 1940s by a Swiss engineer named de Mestral, the name *Velcro* is a blend of the French words *velours* (velvet) and *crochet* (hook). Nowadays, it can be found on everything from sneakers to spacesuits. You can buy it in different widths, colors, and strengths. While it may be tempting to use it for every strap attachment you need to make, you will have to allow for its limitations.

Buckles and Related Hardware

When you need a quick way to disconnect your straps that will still reliably hold your parts together, there are now countless options available in your local fabric shops or from numerous online retailers. Figure 14-5 shows some examples of fairly common plastic strap hardware. Most of these are available in metal, as well.

PARACHUTE BUCKLES   Also known as *side-release buckles*, these are commonly found on luggage, sporting equipment, and tote bags. They're

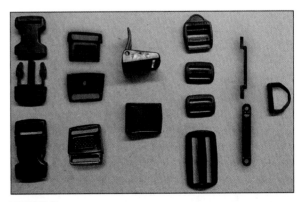

FIGURE 14-5: From left to right, parachute buckles, center release buckles, cam buckles, slide adjusters, footman loops, and a D-ring

two-part, black, plastic buckles that snap together and create a strong hold. Releasing them is a simple matter of squeezing the buttons on the sides together. They can be found in every color and width imaginable, and they can be flat or curved on the backside. There are also different versions that have adjustable connections for straps on one end or both ends.

CENTER RELEASE BUCKLES   These are compact, lightweight buckles usually used in clothing. You'll occasionally see them used on lightweight dog or cat collars or for other lightweight applications. Small, simple, and discreet, center-release buckles are a good option when you want to avoid the bulk of a side-release buckle.

CAM BUCKLES   Commonly found on diving belts, cam buckles are great light-duty straps. They close with a positive lock and do not slip, making them ideal for applications where you want to be able to quickly adjust the tension on a strap without having to deal with lacing and unlacing a regular slide adjuster.

SLIDE ADJUSTERS   Strap adjusters are very versatile solutions to a lot of strapping problems. You'll usually see them built into backpack shoulder straps. They are often used in light-duty utility tie-down straps. Simply lace the webbing through the locking bars, pull the strap tight, and it's good to go.

FOOTMAN LOOPS   These are great pieces of hardware for anchoring a strap to a piece of armor. They are most commonly used on boats, fire trucks, attic doors, and old-fashioned luggage.

D-RINGS   These are used to attach the end of a strap to some other piece of hardware.

How to Attach Straps to Stuff

Once you have your straps and buckles, you'll need to actually find a way to bind them to the other parts of your armor. You're basically looking at two different options: mechanical or chemical.

Nuts, Bolts, Screws, and Rivets: The Mechanical Option

Whether your armor parts are made of foam, paper, or wrought iron, there's no better way to attach your straps than to use an actual mechanical connection. If there are visible fastener details on the surface of your armor, you might as well put them to practical use and attach straps to them, as shown in Figure 14-6.

Pop rivets come in handy for these kinds of applications. Just be sure to use washers wherever possible. This will increase the surface area that is captured by the rivets, making them harder to pull through the armor while trapping the straps in place securely, as shown in Figure 14-7.

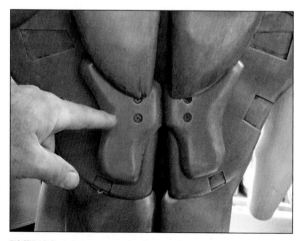

FIGURE 14-6: Visible rivet details on Hunter's backside make for convenient places to attach straps as securely as possible.

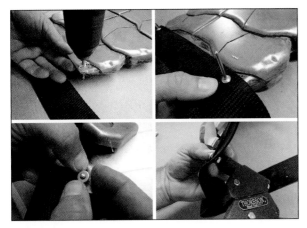

FIGURE 14-7: Installing a strap using a pop rivet and washer

If you are able to use regular nuts and bolts, just remember to cut off the extra on the inside so that it doesn't dig into you while you're walking around. That's fun for nobody.

Finally, if your armor has some thickness to it or there's a cast resin detail piece on the surface, you can use regular wood screws to attach straps to the inside of the armor. Again, it's a good idea to use a washer to help trap the straps in place, as shown in Figure 14-9.

Glues and Goos: The Chemical Option

If there's no easy way to attach the straps with hardware, you'll just be stuck using some kind of glue. It's rarely as reliable as a mechanical connection, but if you use the right kind of adhesive, there's a good chance everything will work out just fine.

Gluing Straps to Styrene, ABS, and Other Sheet Plastics

Since Hunter's armor is made mostly of vacformed styrene, the straps are simply glued in place using a cyanoacrylate adhesive (CA; i.e., Superglue), as shown in Figure 14-10. This provides a decent mechanical bond with the styrene as long as the surface to be glued is scuffed slightly with a piece of sandpaper first. It also has to soak into the nylon

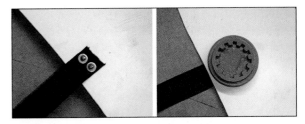

FIGURE 14-9: Small screws with washers (left) can be a great option for holding down straps when they can screw into a cast detail piece on the outside of thin plastic armor (right).

webbing in order to mechanically bond the fibers to the sheet plastic surface.

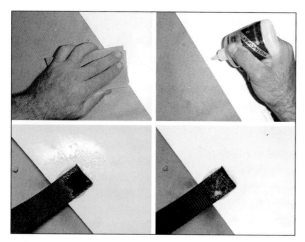

FIGURE 14-10: Gluing nylon straps to a styrene sheet with CA adhesive

Another effective option is to go ahead and rivet the strap onto a scrap of the same type of plastic sheet that your armor was made of, then solvent weld it into place, as shown in Figure 14-11. This way, you'll have a nice mechanical bond to hold the strap onto the plastic, but you won't need to have a visible fastener on the surface.

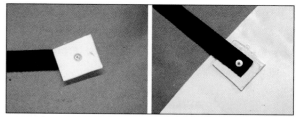

FIGURE 14-11: After riveting a piece of scrap plastic onto the end of a strap, that scrap can now be securely glued to the inside of the plastic armor using solvent adhesive.

How Does Glue Work?

Suppose you're trying to stick two things together—let's call them *Thing One* and *Thing Two*. You'll need to stick them together with Thing Glue. You must consider two different factors: adhesion and cohesion. *Adhesion* is what holds Thing Glue to Thing One and Thing Two. Cohesion is what holds Thing Glue together so it doesn't just crumble or evaporate in the middle. The glue needs an adhesive property to stick to things, but if the glue itself is weak (lacking in cohesion), it doesn't matter how well it sticks because the resulting bond between the parts will break and you'll just be left with two separate pieces with glue stuck to them.

Different types of glues will work in different ways. The simplest adhesives require two slightly rough surfaces. The roughened surface will have tiny pits (often referred to as *tooth* because it allows it to bite onto the glue) so when you add a liquid that can somehow harden between the two surfaces, it will fill in the pits on the surfaces and mechanically bind them together.

Other glues actually make molecular connections to both surfaces, which is called a *chemical bond*. These bonds tend to be stronger and more reliable than mechanical bonds, since the two surfaces being bonded together effectively become one cohesive object.

When selecting a glue, you need to make sure that it will be able to bond to both of the pieces that are being glued together. This can be a challenge when you're gluing together two different materials. As with all things, don't hesitate to experiment to find what works best for you.

Gluing Straps to Fiberglass and Rondo

If your costume is made out of reinforced Pepakura parts, you're going to have to adhere the straps to the bare fiberglass or Rondo on the inside. There are a few ways to go about this. First, you can mix up a batch of the same material and just embed the strap right into the structure of the part. Just be sure to soak the webbing all the way through with the resin in order to ensure a strong mechanical bond.

Second, you can use a compatible epoxy adhesive to glue the straps into place. Just bear in mind that not all epoxies are created equal. As a general rule, the longer they take to cure, the stronger the resultant bond will be. You should also keep in mind that some brands are better than others and most offer different formulas that will work better or worse depending on the materials that you're bonding together.

As with all things, give yourself enough time to experiment and find out what will work best for your specific application.

Gluing Straps to EVA Foam

Getting your straps to stick to EVA foam can be difficult. Ideally, you'll want to give the inside surface a bit more tooth so that the glue has something to help it hold on. Just follow the six steps shown in Figure 14-12.

1. Trace out the end of it on your armor piece with a permanent marker.

2. Score the area with a sharp knife.

3. Apply heat to open the scored lines and seal the surface.

4. Apply hot glue to the scored area.

5. Carefully place the strap and press firmly.

6. Once that cools a bit, you can add a bit more hot glue around the outside of the strap to reinforce the bond.

If you need to attach Velcro to a piece of EVA foam, you may find that the hot glue doesn't get much of a grip on the smooth backside of the Velcro. To work around this, just sew a piece of nylon webbing onto the backside of the Velcro. That'll provide a surface that the hot glue can grab onto.

Design Considerations for Strapping Your Armor Together

If you've ever gone on a long backpacking trip, you'll know that proper load distribution can be the difference between a comfortable stroll with some

FIGURE 14-12: Gluing straps to EVA foam

handy gear, and an agonizing forced march with a massive burden gnawing its way into your soul. If you hang everything off your shoulders from a couple of pieces of thin web strapping, you're on the fast track toward discomfort, pain, suffering, and misery. First you'll notice some mild pain in your shoulders where the straps dig in or chafe. Then, your spine will begin to ache as it hefts the bulk of the load. Finally, the soreness will linger long after you've taken off the armor, even after the pinched skin and tired muscles recover.

Instead, try to set up your strapping arrangement so that any heavy loads are carried mostly around the waist and hips.

Ideally, parts that need to hang on your body should be attached to the lowest possible point.

Hang your thigh armor so that most of the weight is supported at the waist—not the shoulders. While it might be tempting to run a pair of suspenders up to your shoulders to hold up the thighs, the better option is to attach them to a snug belt or load-bearing waist cincher of some sort.

Torso armor will often have no alternative but to hang from your shoulders. It's still a good idea to minimize the weight that you mount to the chest and back armor. If at all possible, hang your shoulder and upper arm pieces from straps that run across your chest and back. This helps to minimize the weight at any one spot.

For any place where a strap applies weight to your body, use the widest possible straps you can get away with. Thinner straps will not spread the load, and eventually they'll start to dig into your tender flesh and begin the agony you'll be whining about for days afterward. At the same time, the wider the straps are, the more likely they will be to restrict your mobility. Wider straps are also a bit harder to adjust.

As a general rule, it's often a good idea to use 1″ webbing wherever possible and 2″ webbing wherever weight is going to be applied to your body. Buckles and fittings are readily available in both sizes, and they are large enough to adjust while wearing gloves.

The Tricky Bits

Science fiction and fantasy armor is usually designed to look good. Sadly, the same awesome designs that you know and love often fail to allow for complete functionality of the joints. Video game designs are notorious for this, and now that computer animation has become so commonplace in movies, impractical armor designs are becoming more and more prevalent. After all, in the middle of an exciting, high-speed, flying fight scene, the audience isn't too likely to notice when some of the pixels start to pass through each other in physically impossible ways.

Since the creators no longer have to take into account whether an actor can perform in his or her costume, it's up to you to make their impossible vision into something that can actually move. You may have to compromise the appearance in favor of making something that's actually wearable.

Here are a few things to look out for . . .

Tight Fits: Wrists, Ankles, and Necks

In order to keep characters looking proportionate, a lot of designers will make armor parts follow the general curves and shapes of the human body. This is usually a good thing. But most people's hands are bigger than their wrists. So sooner or later you'll find yourself trying to build something that looks just right and find out that you've got a pair of gauntlets with wrists so small that no human could ever squeeze his or her hand through.

If you're the kind of person who has feet, you'll run into this exact same problem when it comes time to make any kind of lower leg armor that wraps all the way around the ankle. It also turns out that most people's heads are bigger than their necks. So if you've made a form-fitting helmet, it may well have a neck hole that's too small to fit over the head it's designed for.

There are four basic ways to deal with this. All four can be modified and used for forearms, shins, and helmets, but we'll focus on gauntlets for now.

The first option is to find a good place where they can be split down their entire length. Ideally this will be a seam that's visible in the design. After gluing in some strips of elastic to act as a hinge along one side, it's just a matter of installing strips of Velcro along the edge of the seam on the open side (Figure 14-13).

To put it on, just slide an arm into the open gauntlet, hinge it closed, and push your arm against the inside in order to push the Velcro together (Figure 14-14).

🔫 Maker Note

The side that faces in toward the wearer is the soft "loop" side. That way it's more comfortable and less likely to snag whatever clothing is being worn under the armor.

After pressing the Velcro together, you'll have a nice, snug wrist opening that's too small for a hand to slide through (Figure 14-15).

FIGURE 14-13: A gauntlet split along a visible seam with elastic and Velcro installed

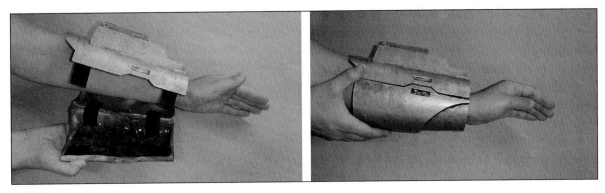

FIGURE 14-14: Sticking the two halves together around an arm

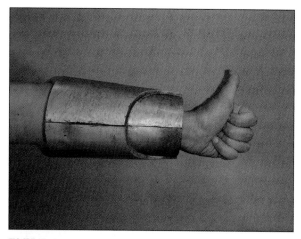

FIGURE 14-15: The nice, snug gauntlet

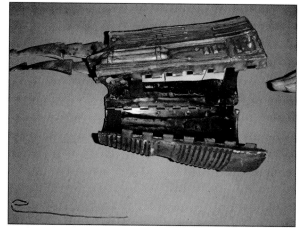

FIGURE 14-16: Hinged gauntlet

There's a second option for an even sturdier connection between the two halves. If you install a length of piano hinge to one side or the other, it will create a sort of clamshell assembly, as shown in Figure 14-16.

The open side can be fitted with Velcro, or better yet, another length of piano hinge with the pin removed. When it comes time to put the armor on, simply lay your arm into the gauntlet, hinge it closed, then stick the Velcro together or slide the pin back into the piano hinge, as shown in Figure 14-17.

The third option is to find a section where the armor has a visible seam where you can make a removable door panel, as shown in Figure 14-18.

Slide your hand through the gauntlet while the panel is removed, then just slip the part back into place. You can use magnets, Velcro, or elastic to hold the removable panel in place once your hand is through. Or, if you build the removable part to fit snugly into place, you can just rely on friction to hold it together.

The fourth option is to make the wrist opening so that it's barely big enough to squeeze your

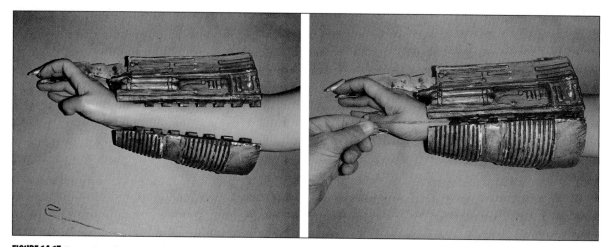

FIGURE 14-17: Fastening the open side of the gauntlet with the hinge pin

bare hand through, then put on bulky gloves that will make your hands appear larger and keep the gauntlet from sliding down off your hand, as shown in Figure 14-19.

Attaching Handplates

The simplest way to attach your armored handplates is simply to glue or rivet them directly onto whatever gloves will be worn with the costume.

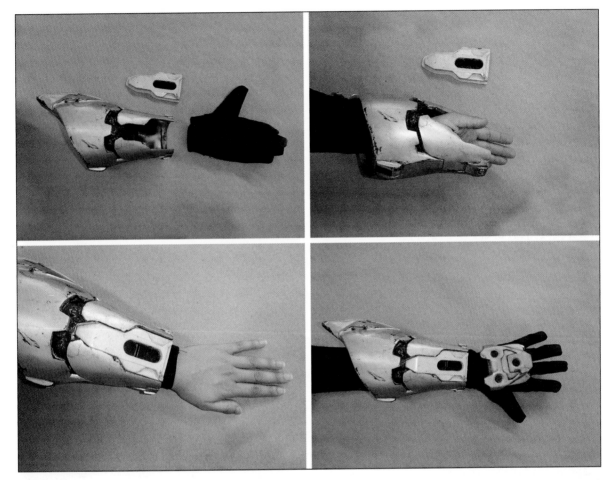

FIGURE 14-18: Hidden doorway

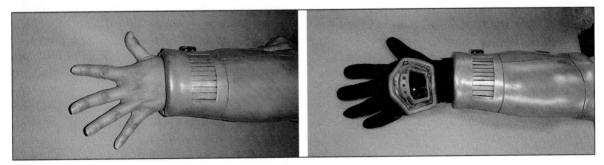

FIGURE 14-19: Gloves to fake larger hands? Why not?

If you want to make them removable so you can use the gloves with more than one costume (or, as in Hunter's case, to change the batteries in any embedded electronics), it's time to get a bit more creative. The easiest thing is to just run a single strap that will cross the palm of your hand and keep the handplate snug on the back of your hand. If this doesn't keep things properly placed, you can use two straps, one around the wrist and one around the palm. This is what was used for Hunter's hand-plates, as shown in Figure 14-20.

If you want something more subtle to hold the handplate in place than a big, wide strap of elastic across the palm, you can also get away with having a strap around the wrist and a small loop of thin-ner elastic or cord that will go around one or more fingers.

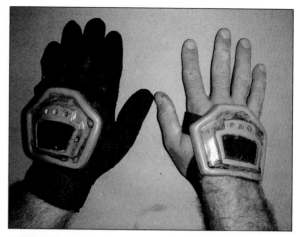

FIGURE 14-20: Hunter's handplates are held in place over the gloves with two straps of elastic.

Attaching Shoe Covers

Once again, the simplest way to attach an armor plate to your shoes is to glue or (better yet) rivet them directly onto the shoes you'll wear with the costume.

If you need to make them removable in order to get the shoes on and off, you can install a set of straps that will go around the sole of the shoe under the arch of the foot, as shown in Figure 14-21.

Knees and Elbows

For armor with separate plates on the knees and elbows, you can usually just have a simple strap on the inside of the joint in order to keep things in place. If you slide your arm or leg into these pieces, all you need is a snug-fitting piece of elastic. If you'll need to make them more adjustable and detachable, you can use Velcro to make them removable, adjustable, and much more comfortable, as shown in Figure 14-22.

FIGURE 14-21: Stirrup-style straps keep Hunter's foot armor in place.

FIGURE 14-22: Velcro: the way to go for comfy knees.

Hips and Shoulders

When it comes to designing a strap harness that allows for maximum mobility, most of your headaches will be found in the areas of the hips and shoulders, where the body's ball-and-socket joints have a much wider range of motion than you find in the hinge joints of the knees and elbows. The best way to allow for this movement is to use a single strap to keep each shoulder or thigh piece from falling down, and another around the arm or leg to hold it down.

Take a look at Hunter's shoulder armor in Figure 14-23. There's a single strap attaching the very lightweight shoulder armor to the strap that connects the chest and back. In order to keep the shoulder from flapping around in the wind, a single piece of elastic is permanently attached inside the front of the shoulder and wraps around the wearer's upper arm. Velcro holds the elastic in place at the back of the shoulder in order to make it easier to get the armor on and off.

Since Hunter's thigh armor wraps all the way around his legs, they're only going to need straps to hold them up. In Figure 14-24 you can see the parachute buckles installed on top of the thighs. These will clip onto the straps that hold the thigh armor up. Capturing these straps under the belt should allow the hips to move without causing the thigh armor to rotate around on the wearer's leg. If they were simply plates that sat on one side of his thighs, they'd also need straps to hold them around the thighs, just like the shoulder plates.

FIGURE 14-24: Parachute buckles installed at the top of Hunter's thigh armor.

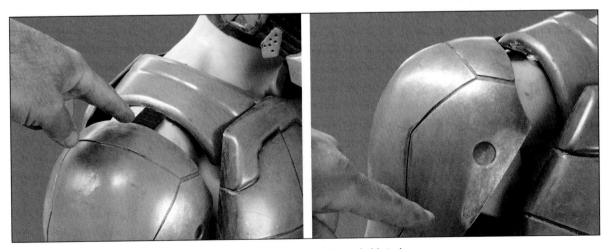

FIGURE 14-23: A single strap holds Hunter's shoulder up while a strip of elastic holds it down.

Buttplates, Codpieces, and Anything Else That'll Keep You from Using the Bathroom in a Hurry

When you're working out a way to strap your armor together, you need to take a bit of time to consider how you're going to answer the call of nature. This can be the difference between having fun in your costume all day, and suddenly having to do an emergency load of laundry.

In Hunter's case, there's no groin armor, so it's easy. But if your character does have a hard plate covering their delicate bits, you'll want to attach it with snaps or Velcro so it can be removed without getting out of the entire ensemble. Also, if there's a plate that hangs down in the front or back, it's a good idea to run a "thong strap" connecting them through your legs from front to back so they don't flop around if you're moving in a hurry. A knight in shining armor tends to be a bit less dashing if his codpiece is flopping around and smacking him in the junk when he walks.

Odds and Ends: A Few More Strapping Tips

1. Once you've figured out the correct length for your straps, cut off the extra, but not all of it. A few extra inches will allow you to retain some adjustability, but cut too much and you'll have useless little pennants start poking out when you move. Add some slides, as shown in Figure 14-25, in order to keep everything neatly tucked in place.

2. When you cut your straps to length, they'll want to fray. If you're using nylon webbing, you can heat-seal the edges by holding them near an open flame (such as a cigarette lighter) for a second or two. If you're going to be doing a lot of this kind of work, consider investing in a hot cutter like the one in Figure 14-26.

3. Buy in bulk. A 5′ roll of 1″ webbing will cost a few dollars at the local fabric store. The same material can be had for a few cents per foot on websites like strapworks.com. With prices like that, you might as well get some extra. You'll use it sooner or later.

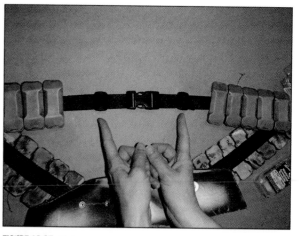

FIGURE 14-25: Slides are a key part of keeping straps in place.

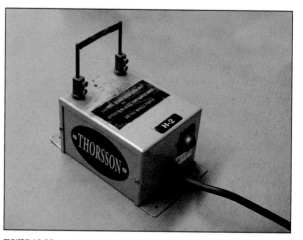

FIGURE 14-26: A hot cutter uses an electrically heated blade to simultaneously cut and seal synthetic straps and rope.

4. Try to permanently strap together as many pieces as you can. Strapping multiple pieces together into sub-assemblies will make it easier to get in and out of your armor, as well as making the whole ensemble stronger in the process. In Hunter's case, the chest and abdomen are strapped together into one large piece, while the back and shoulders are combined into another integrated piece, which is shown in Figure 14-27. Once the back assembly is buckled to the chest assembly, this whole arrangement can be donned much like a shirt.

5. Experiment. What works for someone else may or may not work for you. Don't be afraid to revisit your plans and rearrange your straps after you've had a chance to try your rigging in action.

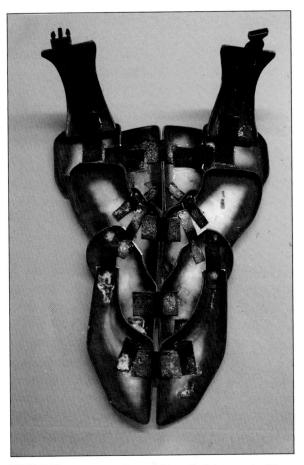

FIGURE 14-27: By permanently attaching all of these parts, it's also harder to lose an important piece.

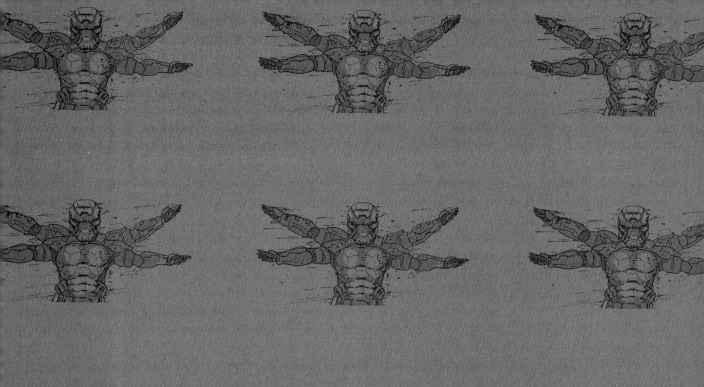

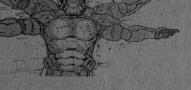

SHOWING OFF

Now That You've Got It, Flaunt It!

NOW THAT EVERYTHING IS built, it's time to take it out into the world. But buckling on a broadsword and boarding a bus is probably not the best plan. Instead, it's going to take a bit of forethought. There are a few things to take along.

Handlers

It's dangerous to go alone . . .

. . . okay, probably not. Regardless, this is one of the many times in life where, no matter how nice the neighborhood where you're marching around, it's a good idea to have a friend along. The main drawback to being completely encased in some sort of expensive, labor-intensive suit of awesome is that it'll usually be at least some kind of cumbersome. More often than not, costume armor pieces will limit the wearer's range of motion, visibility, comfort, and endurance. In these cases, it's good idea to bring one or more "handlers."

Handlers can help with the following:

HARASSMENT MITIGATION   For some reason there will always be at least one obnoxious little kid who feels compelled to run up and kick a costumed character in the shin or punch them in the groin (Figure 15-1). Having a handler watching your back is invaluable.

CHILD SAFETY   When you can't see, it's nice to have someone steer you away from stepping on the little kids who *aren't* punching you in the groin.

CARGO   Want to bring water? A wallet? Cellphone? Lifesaving medications? If your costume doesn't include pockets, the handler can handle it. Otherwise you'll have to tuck all of that into your underwear.

DOORS AND STAIRS   If your armor is really bulky, the simple acts like opening a door or going up a flight of stairs can be problematic. Having a shoulder to lean on or a pair of ungloved hands to push the elevator buttons will help to hide the fact that your massive armored gauntlets make you completely helpless.

CROWD CONTROL   A friendly, assertive handler can organize a queue for folks who want to pose for pictures with you, as well as scoot people out of your path when you need to find some place to take a break.

DAMAGE CONTROL   Murphy's Law will be in full effect at all times. The handler can help by picking up any parts that fall off and patching up any wardrobe malfunctions on the go.

Field Repair Kit

No matter how well a costume is built, sooner or later something will go wrong. Knowing this ahead of time, you can be prepared by packing up a small pouch, like the one shown in Figure 15-2, with a few essential items so you can make repairs wherever you are and get back to trooping around as quickly as possible.

Here are just a few essential items to pack around:

- Cyanoacrylate adhesive and accelerator
- Gaffer tape or duct tape in a color that will blend with your armor

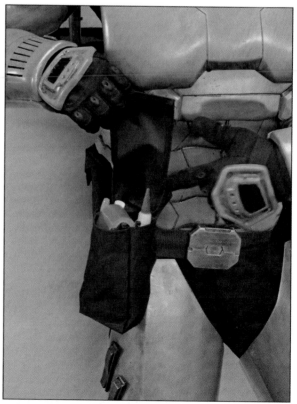

FIGURE 15-1: Kids will be kids. Like it or not.

FIGURE 15-2: A field repair kit packed neatly into one of Hunter's belt pouches

- Safety pins
- Zip ties
- Needle and thread
- Paint markers to touch up any scratches in a hurry

Posing for Photos

Chances are, your character probably has at least one or two iconic poses that show up in the comic, game, or show that it originally came from. It's a good idea to get completely dressed in the armor and practice these poses a few times in front of a mirror (Figure 15-3). Remember, what

Maker Note

Instead of carrying around paint in every color you have on your armor, it's usually a good idea to carry around markers that are black and silver so you can turn that new scratch on your armor into a bit of battle damage. That way it doesn't look like an accident anymore.

feels awesome may look a little silly and what looks good is often borderline torture when you're the one in the suit.

FIGURE 15-3: Practicing the Hunter's pose from the cover of the comic.

Have some fun with it. Also, try to come up with a handful of different poses so you can change it up a bit while you're out and about. That way you won't be disappointed the next day when you see countless photos of yourself on social media in the exact same pose over and over and realize that you might as well be a cardboard cutout (Figure 15-4).

Finally, when someone is snapping away, make sure you know where the camera is. There's nothing worse than making yourself look silly by being the only person looking left when everyone else is looking straight (Figure 15-5).

Planned Photoshoots

Eventually, you're going to want to get some better photos of the finished costume to hang on the wall in your office, or to show your future grandchildren how cool you used to be. So get yourself a photographer who knows what they're doing and start snapping.

Location, Location, Location!

While a studio photoshoot with a plain, monochromatic backdrop can do a nice job of showcasing a finished costume, there's a lot to be said for photos that look like they were captured in the wild somewhere. But to get a truly natural-looking photograph, there are a lot of things that will have to be faked (Figure 15-6).

While building your armor, keep an eye out for shooting locations during your daily travels. Here are just a few things to look for in a location:

APPROPRIATE BACKGROUND   Is your character supposed to be living on some futuristic world

FIGURE 15-4: Seriously. Grow a few more poses.

FIGURE 15-5: Get with the program.

FIGURE 15-6: Hunter, perched high on a rooftop surveying his next target, is actually standing on the edge of a low retaining wall while the photographer lies on the ground in front of it. Photo by Shawn Thorsson

populated entirely by sharp rocks and glass skyscrapers? Having a Tudor-style house in the frame might detract from the illusion a bit. Is it some sort of snowbound wolf warrior woman* who lives in the wild woods of wherever? Find some snow and some woods and make it work (Figure 15-7).

ACCESSIBILITY   If the armor is difficult to walk in, it'll be pretty hard to hike it down a long, slippery trail along the edge of a cliff above the place of your untimely death. Maybe you should find someplace closer to the parking lot.

PERMISSION   A word about trespassing: don't.

LIGHTING   While it's possible to bring along portable artificial light sources, natural light is usually better. Plus it's free. Renewable, too.

Finally, make sure you give your photographer proper credit, and get his or her permission to use the images before you post them online, start selling prints, or publish them in a book or anything of that sort.

---

*What if her name was *Accalia*? It's Latin for *She-Wolf*. That's a winner all the way. We're running with it.

FIGURE 15-7: Snow? Woods? Possibly the glow of a burning village off in the distant background? This location has everything!

Photo by Mandy Valin

INDEX